Telecommunications

A software professional's guide

Clive Tomlinson

Addison-Wesley

An Imprint of Pearson Education

Harlow, England • London • New York • Reading, Massachusetts
San Francisco • Toronto • Don Mills, Ontario • Sydney
Tokyo • Singapore • Hong Kong • Seoul • Taipei • Cape Town
Madrid • Mexico City • Amsterdam • Munich • Paris • Milan

PEARSON EDUCATION LIMITED

Head Office:
Edinburgh Gate
Harlow
Essex CM20 2JE
Tel: +44 (0)1279 623623
Fax: +44 (0)1279 431059

London Office:
128 Long Acre
London WC2E 9AN
Tel: +44 (0)20 7447 2000
Fax: +44 (0)20 7240 5771
Website: www.aw.com/cseng

First published in Great Britain 2000

ISBN 0-201-67473-4

British Library Cataloguing in Publication Data
A CIP catalogue record for this book can be obtained from the British Library.

Library of Congress Cataloging in Publication Data
Applied for.

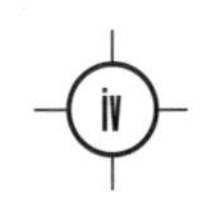

The Publishers wish to thank Objective Systems Integrators for permission to reproduce figures 11.6, 11.7 and 11.8, and BTCellnet for permission to reproduce figures 11.6 and 11.7.

10 9 8 7 6 5 4 3 2 1

Typeset by M Rules, London
Printed and bound in Great Britain by Biddles Ltd, Guildford and Kings Lynn

The Publishers' policy is to use paper manufactured from sustainable forests.

Telecommunications

A software professional's guide

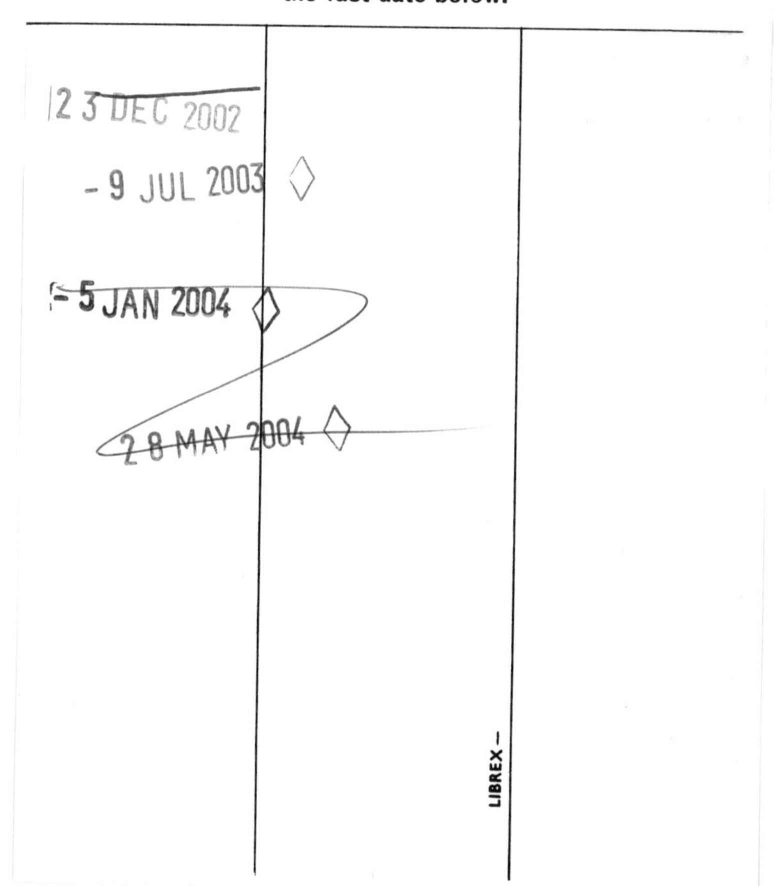

For my girl Annie

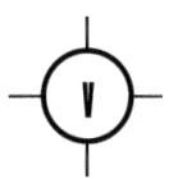

Contents

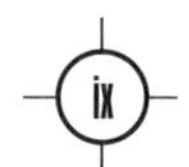
ix

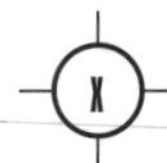
x

About the author

Clive Tomlinson has worked for over twenty years as a telecoms software engineer and systems architect. He has developed, supported and managed software technology for packet mode data networks, circuit switched voice networks, and their supporting management systems.

Having worked with many of the world's largest telecoms network operators, equipment vendors and software companies, Clive is well able to give a lucid high-level account of telecoms technologies, the structure of the industry, and the special software methods that it uses.

Clive is a practising telecoms software consultant, a popular speaker at technical conferences, and the author and presenter of a number of telecoms training courses. He is a chartered engineer, a member of the IEE, and a smashing bass player.

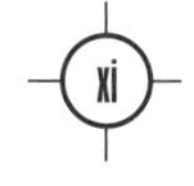

Acknowledgements

Many people have helped me with this book.

Most of the content is due to the many telecomms engineers who explained things to me over the years, and the many software engineers who made me straighten out my thoughts by explaining things to them. I thank them all, and in particular:

- The very obliging engineers (Dave Huxford, Graham Monro, Malcolm Murphy, Rinaldo Tempo, Harry Tomlinson and Paul Tomlinson), who reviewed my efforts and contributed from their own experience.
- Gavin Duckett and his colleagues at BT Cellnet, who kindly provided the screen shots for Chapter 11.
- The staff at Pearson Education, who encouraged me to get writing, and the directors of IPL Information Processing Limited (my daytime employers), who gave the work their support.

Most of all, my thanks go to Annie, Mark, Guy and Amy Tomlinson, who were very patient.

Introduction

This book is intended to help software professionals to practise in the telecommunications industry. Telecoms is technologically dominated by software and is therefore an industry that is acutely hungry for the services of capable software people. However, if someone with only a general IT background goes into a telecoms company, what he hears is often baffling, as Figure 1 shows.

This may strike the software generalist with awe, which would be a pity, because actually the particular ideas in the picture are

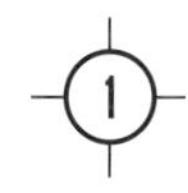

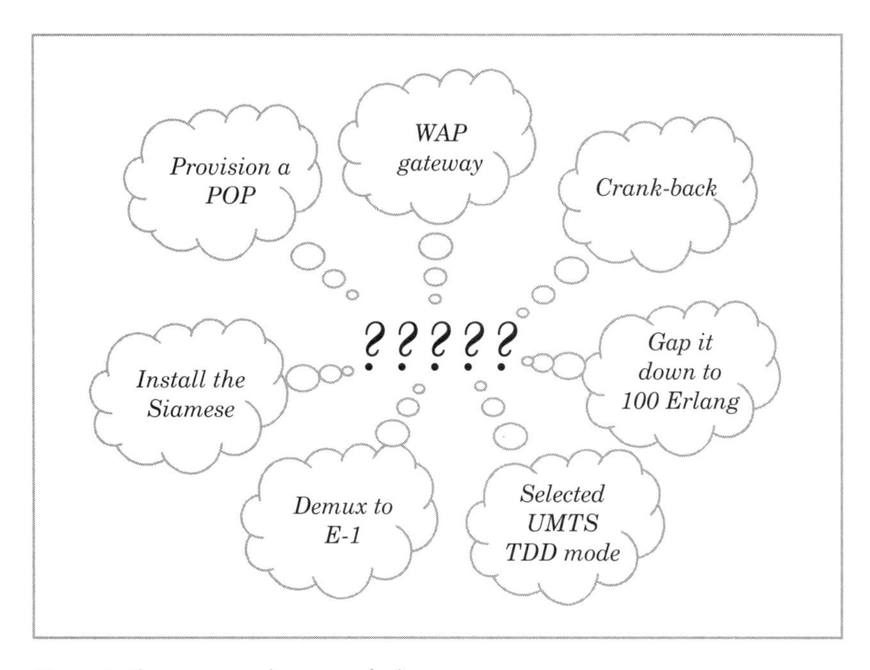

Figure 1 The mysterious language of telecoms engineers

quite simple. But there is a genuine obstacle to be overcome. Telecoms networks involve a truly mind-boggling array of interoperating technologies. The field is so laden with jargon that unless the software person has an understanding of a wide range of technologies, he or she has little hope of communicating with the telecoms engineer.

Consider the example of Figure 1. To make sense of that requires a grasp of switch-routing heuristics, internet mobility, cable TV networks, third-generation cellular networks, transmission systems, provisioning and traffic management.

Happily, a grasp is all that is needed. To work effectively with telecoms engineers, you do not need to know their job anything like as thoroughly as *they* do. Some parts of telecoms technology are particularly software-intensive, and in those parts, the software professional may need something more than a superficial understanding. But by and large, a little well-selected knowledge can go a long way.

The aim of this book is to give the reader a wide-ranging understanding of the telecoms industry, right the way from the top-level business processes, through the network technologies, down to the special techniques used in telecoms software. Armed with that understanding, a capable software person should be able to communicate intelligently with telecoms engineers.

This book does not replace the many complete and detailed specialist technical texts available. This is a generalist work. It is deliberately light on detail, but it covers most of the areas with which a telecoms software engineer is likely to have to contend. Once this book has got you started, you can go on, if you will, to immerse yourself in the detailed technical works. I am a software engineer, and I assume that my reader is one also. But a software background of undergraduate level is more than enough to make sense of what follows.

Finding your way around

Telecoms is a complex tangle of technologies and businesses, and to make it understandable we have to cut it up into easy chunks. Figure 2 shows how this book carves up the subject. It is just one of many possible ways of doing this, but I have chosen it to present the material in a sequence which will be least confusing for the reader.

To get an appreciation of why telecoms engineering is such a lively business, we start in Chapter 1 by looking at the telecoms market: who the players are, how they get their revenues, what their goals and challenges are, and so on. This is a technical book, not a market analysis, but we need to share some idea of what is going on in the market so that we can appreciate the significance of the technologies on which this book focuses.

Chapter 2 introduces some architectural models to help you understand the structure of telecoms networks, and some basic transmission principles.

Chapters 3, 4 and 5 are a Cook's tour of transmission systems, starting with fixed bit transports such as optical fibre, then mobile technologies like UMTS, and ending up with the complex multiplexing systems which make the bit transports usable.

Chapters 5 and 6 are a similar tour of circuit mode switching systems, starting with the signalling systems that define the protocols to make the bit transports meaningful, then looking at the more ordinary telecoms switches (telephone exchanges), and lastly the *intelligent networks* switching technology.

Chapters 8 and 9 explain the IP data communications technologies that support the Internet. The applications of these technologies are then reviewed, in particular the World Wide Web and internet telephony.

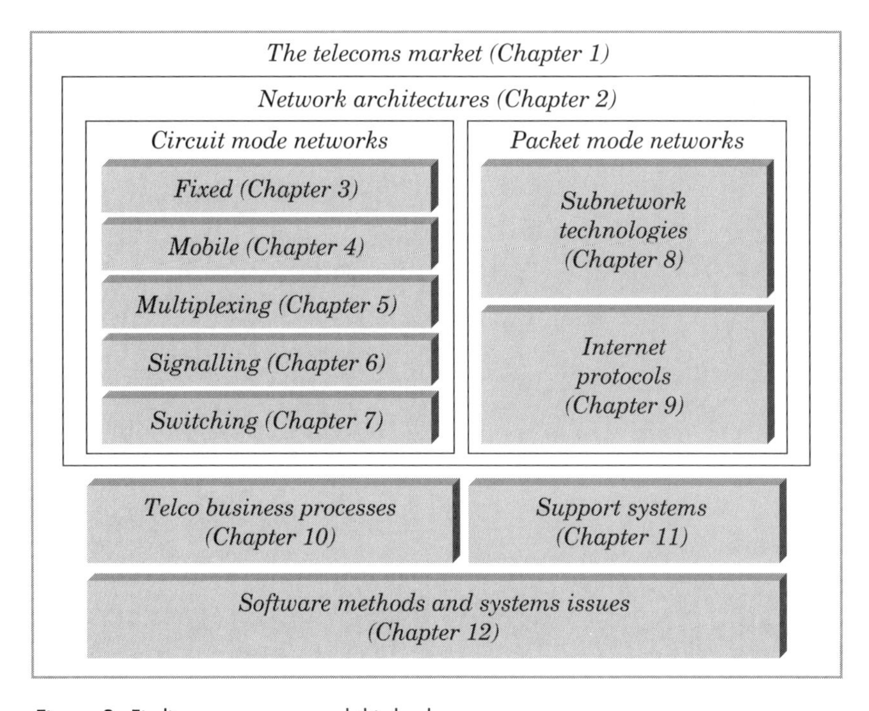

Figure 2 Finding your way around this book

Chapters 10 and 11 then explain the kind of business processes that go on in a telco, and the computing systems that they need to support them.

Lastly, Chapter 12 discusses the half dozen software techniques that have evolved to suit the telecoms industry's special needs.

1 The market

The telecommunications market is immense and fast-moving. Corporations and technology fads come and go. There is, however, a fairly stable high-level structure to the market, which persists even though the names and the technologies change. This chapter starts by telling the story of how the market came to its current state, then presents that high-level structure and relates it to current market conditions and trends.

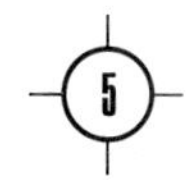

But what is telecommunications?

'Telecommunications' is a much-abbreviated word. It is more usually called 'telecoms', 'telecom', or 'telecomms', depending on where you come from or how hip you wish to sound.

Part of its meaning can be got from its classical roots. *Communication* comes from the Latin word for *sharing*. *Tele* comes from the Greek word for *far away*. Telecommunications is about enabling people (or machines) that are far away from each other to share information.

The other side to its meaning is a legalistic one. Telecommunications is about communication in a *three-party relationship*, where two distinct legal entities are enabled to communicate, through the work of a third-party network operator. So, telecommunications is meant to be about network operators helping other legal entities to share information over large distances.

However, even this simple definition is not quite reliable. For example, enterprise networking now often uses the technologies and business models of long-distance telecommunications within a single legal entity. Or the communicating parties may sometimes be just next door to each other. The blurring of distinctions and lack of precise terms are just symptoms of the rate at which the telecoms business is developing. Language may catch up with it sometime.

Technological development

The telecoms market has developed over about 150 years. The story, on the whole, has been of a hungry market seizing voraciously on each new technological development. There have been a few exceptions to this rule,[1] but generally speaking, availability of technology has been the primary controlling factor in the market.

The dynamics of the market have been shaped by the capacity of the systems available. Initially the market was led by public telegraphs. These were usually point to point systems, with little in the way of a network, and with no requirement for the end users to invest in special equipment. Their utility was therefore roughly proportional to the number of users, and so traffic (and revenues) tended to increase more or less linearly.

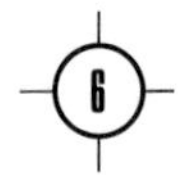

Over most of the last 100 years, telephony has dominated the market. Telephony requires investment in telephones at the customer's premises, and needs a switching network providing high levels of interconnection between large numbers of end points. This interconnection means that the utility of telephony networks is roughly proportional to the square of the network size, at least for communities of interest. A network of just two users can offer only one possible connection. One with three users can offer six, and so on. Consequently, the growth of telephony networks has been more or less exponential. By the late 1930s, some 50 years after practical telephones appeared, very few ordinary people had home phones. But after another 50 years, the home without a phone was becoming a rarity, and homes with two or more phone lines were commonplace.

Over most of the past 100 years, telephony has dominated the market.

Data services are expected to dominate the market in the 21st century. The growth of data networks is not even limited

[1] For example, Videotex, N-ISDN and satellite personal mobile systems, which have all met with only patchy success. These exceptions are the more notable because they are so few.

by the number of users, because the traffic is between machines, and much of it is without the users' direct involvement. The growth of data communications services continues to defy prediction; the best-reasoned forecasts are generally feeble underestimates, and of as little value as the wildest guesses. But more about market trends later. Let us go right back to the beginning.

Before specialist telecommunications organizations were invented, the communicating parties were responsible for their own transmission. Speech is a powerful system which has survived for 5000 years or more but is restricted to very short distances. Long-distance communication was possible only by travelling oneself, or by using the memory of a private messenger, but this was unreliable and too costly for use outside government or big business.

Writing offered portable digital message storage, and thus vastly increased the capacity, security and reliability of the messenger system. Other early forms of digital transmission were either medium-range optical systems, such as smoke signals, naval flags and semaphore, or short-range audio, as in the battlefield command system of trumpet calls.

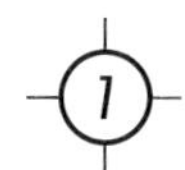

Many nations went on to develop long-haul optical communication systems using repeater stations on hilltops. These began as one-bit capacity beacon chains, providing early but non-specific warning of grave danger. By the late 1700s there were a number of optical telegraph systems, where each hilltop repeater station had a tall wooden frame carrying semaphore arms, or a number of panels which could each be made to show a dark or a light aspect. Optical telegraph systems could send complex messages, at a data rate of several bits per minute, at a speed far greater than any form of transport then available.

All of these systems were, however, so costly that they could be justified only for military purposes. They were also often rather restricted in the scope of what they could communicate. For example, the English admiral Horatio Nelson was celebrated in school books for flying the brusque signal 'England expects that every man will do his duty' before the battle of Trafalgar. Actually, he ordered the more encouraging 'England confides . . .'[2] to be transmitted, but the naval flag code of his time had no way of saying that.

[2] i.e. '. . . is confident . . .'

The first telecommunications organizations, in the sense of organizations that offered to effect communication between other parties, were the posts, which offered a kind of packet-switched service, limited to the speed of a horse.

Digital circuit-mode transmission

In 1829, Stephenson developed the first practical steam locomotive, which offered the possibility of increasing postal transmission speed by a decimal order of magnitude. But this was eclipsed by the development, by Wheatstone, Morse and others, of the first electrical digital circuit-mode transmission system, the electric telegraph, in the years 1831–1838. An electric telegraph consisted of a pair of wires, with a switchable current source at one end and a current detector at the other. The fastest available form of signalling was Morse code, between trained operators. This offered near-instantaneous transmission between telegraph offices, at a rate of roughly 10 bps[3] at the hands of an expert operator. Telegraph networks were laid between centres of commerce and government, and the telegraph became the preferred communication system for government and business.

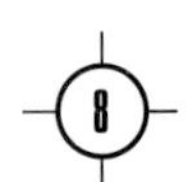

However, the telegraph system was restricted to doing the long-haul part of the communication link; users had to go to their local telegraph office to send or receive messages. In 1844, the telegraph technology was made more accessible through the telegram service, where a lad on a bicycle would deliver a paper transcript of the message to its destination, and take a reply if requested.

In 1850, the first international telegraph link was laid, across the English Channel from Dover to Cap Gris Nez. By 1858, a transatlantic telegraph cable had been laid from Ireland to Newfoundland, although because of technical problems it did not come into regular service until 1865. The world suddenly became smaller.

The data rate achievable over a telegraph using Morse code is limited by the capacity of human operators, not by the electrical technology. To improve on this, in 1874 Baudot launched the first *time division multiplexing* (TDM) system. The sending operator would use a combination of five two-position keys to

[3] Twenty words per minute, using standard abbreviations, was about the limit.

signal a letter or digit as a single operation. Baudot's system took this five-bit code word, and sent the bits down the line sequentially, at a rate much faster than a human could manage. At the receiving end, Baudot's apparatus reconstructed the five-bit code word and marked a paper tape to indicate the character received. A derivative of Baudot's five-bit code persists to this day in the ASCII character code.

Analogue circuit-mode networks

The slightly awkward digital (written) user interface of the telegraph, the need for costly human processing of every message, and the lack of immediate two-way dialogue meant that its use was restricted mostly to the business community. The 'killer application' for electrical circuit mode transmission was the analogue telephone. On March 10 1876, Mr Bell spoke the message 'Mr Watson, come here, I want you' into his rudimentary transmitter, and his assistant at the other end of the house heard him through a similar apparatus. This huge leap in user-friendliness led to a rapid development of local telephone networks.

Initially, telephone networks used human operators to switch calls. Each user of a local network had a pair of wires running from his or her telephone instrument to a 'central office' or 'telephone exchange'. A user wishing to make a call would crank a magneto to attract the operator's attention (via a mechanical 'drop' indicator), then would speak to the operator to request a connection. The operator would connect an alternating current source to the requested line, to make its bell ring. When the called party answered, the operator would use a patch cord to connect the users' lines.

In 1889, Almon Strowger patented the first automatic exchange system. However, calls were still restricted to the local area because the signal power generated by the microphone at one end of the circuit was the only power available to overcome the losses in the network and drive the earphone at the other end. While telegraph networks could be run over long distances by using digital amplifiers based on relays, the analogue electrical amplifiers needed for telephony were not available until 1906 when Lee de Forest invented the triode valve. Valve amplifier technology made long-distance telephony possible, and the telegraph lost most of its remaining advantage.

Long-distance telephony transmission enabled calls between users connected to different exchanges, and so long-distance carrier networks developed to connect the local networks. Telephony addressing became standardized as decimal numbers within each exchange's local network, which enabled the use of automatic switching equipment. But exchanges were addressed by alphabetic names,[4] and inter-exchange calls still had to be routed manually.

Through the first half of the 20th century, analogue circuit mode voice networks spread and developed. By the late 1930s, long-distance transmission of many calls over a single coaxial or balanced-pair trunk was possible, using *frequency division multiple access* (FDMA), as was wireless telephony.

Analogue over digital circuit mode

Analogue voice networks offered rather limited speech quality. Particularly over long distances, where several amplifier stages were traversed, the quality could be very poor, because each amplifier introduced its own distortion and amplified not only the speech but also all the electrical noise picked up along the way. In 1938, Alec Reeves invented *pulse code modulation* (PCM), the first way of using digital transmission to carry voice signals. PCM (as Chapter 2 explains) enabled high-quality, long-distance transmission.

The Second World War distracted much of the world from developing or even properly maintaining its civil telecommunications networks.

The Second World War distracted much of the world from developing or even properly maintaining its civil telecommunications networks, and so by the late 1940s many countries' networks were seriously inadequate. This was in a way convenient, because it enabled the network operators to take advantage of the invention of the transistor by Bardine and Brattain in 1948, which enabled the construction of relatively cheap large-scale networks using PCM and digital transmission. In 1956, the first (analogue) transatlantic telephony cable was laid, using submarine valve repeaters. This was followed by transistor-based equivalents, and then by

[4] So, for example, 'Pennsylvania 65000', popularized in the song by Glenn Miller, was the telephony address of a hotel in New York, which was served by the 'Pennsylvania' central office.

digital systems. By the early 1970s, most long-distance telephony used digital transmission.

Switching was also developing. By the 1950s the routing capabilities of telephony switches were advanced enough to support automatic long-distance call routing,[5] and the text names of exchanges were first reduced to three-letter codes (interpreted through letters printed on the telephone dial) and later to purely numeric codes. In 1972, the first software-controlled (but analogue) telephony switches were installed, and by 1976 fully digital switches were available.

Optical fibre transmission systems were first installed in the late 1970s, and then *wavelength division multiplexing* (WDM) in the late 1980s. By the end of the 20th century, the global circuit switched digital telecommunications network had become by far the largest and most complex machine in the world. Telecommunications technology had become as vital to mankind as oil or electricity. However, by the mid-1990s it was clear that circuit mode networks were being superseded by packet mode systems.

Digital packet mode networks

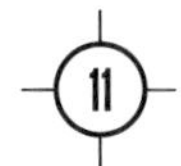

The invention of the electronic digital computer by Tommy Flowers (a telecommunications engineer) and others, for secret code-breaking work in the 1940s, introduced the possibility of packet mode digital data networks. A computer could be connected to a number of digital transmission lines and route messages between them.

In the 1960s, the US Department of Defense put together the first large-scale experimental packet mode network, called the Advanced Research Projects Agency Network (ARPANET). ARPANET was used extensively outside the military for exchanging data among the academic community, and by the 1970s had spread across much of the world. I used it in the late 1970s, and can testify that it was a cranky unreliable affair, usable only by dedicated geeks. But as the alternative to ARPANET was to wait for the postman to deliver a spool of magnetic tape, the network's quirks were easily forgiven.

In the 1980s I worked in the development of JANET, the *joint academic network*, which was a faster, more robust remake of

[5] These were called *subscriber trunk dialling* (STD) in the UK, and *direct distance dialling* (DDD) in the US.

ARPANET, and which became a part of the huge global conglomeration of data networks now known as the Internet.

The Internet was still of interest to academics and tech-heads only, until in 1989 Tim Berners-Lee's proposal for the world wide web (WWW) of documents connected via hyperlinks was first implemented. Just as the telephone had made circuit mode transmission appealing to the masses, so the WWW was the popularizing application for packet mode networks. By 1991, commercial use of the Internet had begun. By 1997 there were an estimated 50 million Internet users, and the loss caused to the world's circuit mode network operators was significant. By 2000, national and global packet mode networks had been constructed, independent of the circuit mode networks.

Commercial and regulatory development

When the telecommunications market began, it was not immediately obvious that it needed any external regulation, and development was largely free and chaotic. As telecommunications services became essential to everyday life and business, state authorities began to intervene to control the availability, quality and cost of service. More recently, there has been a trend to deregulation, allowing telecommunications service providers (telcos[6]) more freedom to follow market demands, within a framework of regulatory checks and public service requirements. At the same time, there has been a removal of the monopolistic positions that many telcos had been allowed to hold. The UK and the US are examples of two nations that followed this model and that often set patterns which the rest of the world has tended to follow.

The UK story . . .

In the UK, the General Post Office (GPO) had a monopoly of the public posts service from the middle of the 17th century, and had acquired a monopoly of the public telegraph service (perceived as not unlike the letter service) from 1869. Initially, telephony was not perceived as a threat, or as much more than

[6] A term originally applied only to network operators in the US, but now used very generally for any company that provides network services, even if it does not actually own a network.

a toy, and so was not subject to regulation. Many local telephone networks were developed, some by local authorities and some by independent companies, of which the largest was the National Telephone Company, which operated dozens of local exchanges. Interconnection of local exchanges was piecemeal. Quality of service, tariffs and technology varied between areas, and long-distance calls had to be made by hopping from one local exchange to the next, across the country. The development of the network was hampered by the operators' need to get legal rights to install wires and poles along public highways, and to negotiate wayleaves with landowners. As these rights could only be gained through individual Acts of Parliament, it was a slow business.

For many reasons, not least the need to improve service quality and establish a national network, the government nationalized the public telephone networks in 1910. The National Telephone Company, all the other independents and most of the local authority telephone system operations were taken over by the GPO. The local authorities, however, were given the option of continuing to run their own networks, subject to licensing by the GPO and conformity to national standards. Several of them did so, but over time, all of them, with the exception of the Kingston-upon-Hull system, were taken over by the GPO.

For much of the 20th century, the Hull system barely kept up with national network standards, despite having an easy territory to serve. However, since the introduction of STD in the late 1960s, the Hull network has flourished, and in the deregulated environment of the late 1990s it has expanded rapidly and profitably.

The pattern of one state-run organization providing both postal and telephone services has been a common one, particularly in Europe. Such an outfit is called a PTT (*posts, telephony and telegraphy*) or a PTO (*posts and telecoms operator*).

Up to 1969, the GPO had not only operating responsibilities but also legislative and administrative ones. It was a part of the UK civil service, headed by a postmaster general, who was a member of the government. In 1969 the GPO was dissolved and replaced by the Post Office, a government-owned public corporation, still encompassing both posts and telecoms businesses. The job of postmaster general was abolished, and the Department of Trade and Industry took over his regulatory and legislative functions.

From the late 1970s, users were allowed to choose and own their equipment, and independent firms were allowed to supply equipment directly to network users. In 1981 the telecoms business was entirely separated from the Post Office, and became known as British Telecommunications (BT). BT retained its monopoly of telecoms services until in 1982, in an attempt to stimulate performance through competition, the government broke BT's monopoly by awarding a network licence (although without any *universal service obligation*) to Mercury Communications, a private company.

In 1984, BT was converted into a public limited company (plc). Initially the government owned 51 per cent of the shares, but three years later it had sold its share holding and completed the privatization of the business. Mercury provided little in the way of serious competition for BT, so the government went on to award licences to more and more network operators. In 1985, mobile network licences were awarded to Cellnet and Vodafone, and in the early 1990s competition in the local loop was stimulated by allowing cable television operators to offer telecommunications services. By the mid 1990s there were more than 100 licensed network operators in the UK, and the government's regulatory body, Oftel, had gained formidable powers to enforce public service obligations and stimulate competition.

. . . and in the US

In the US, within a year of Bell's invention, the Bell Telephone Company was formed. The Bell Company encouraged local entrepreneurs to set up small local networks under licence, all over the US. In the mid 1890s, Bell's basic patents expired, and so it became possible to build a telephone network without a license from the Bell company. A huge number of independent telephone operating companies sprang up. At their peak, there were about 6000 *local exchange carriers* (LECs); even now, following years of consolidation, there are still more than 1000.

The Bell Company encouraged local entrepreneurs to set up small local networks under licence, all over the US.

The Bell Company, which by then was called the American Telephone & Telegraph Company (AT&T), had the advantage of having constructed a national long distance network during the years of its effective monopoly. AT&T was not co-operative about letting the *independent LECs* (ILECs) connect to this network, and instead made a habit of buying them up. This was perceived as anticompetitive, and in 1913 the US Justice Department pushed AT&T into making an out-of-court settlement called the Kingsbury Commitment, wherein AT&T agreed to provide interconnect services to all LECs, and to submit any further proposed takeovers to the Justice Department for approval.

Regulation of the telcos was done independently by each state's legislature, although in 1934 the Federal Communications Commission (FCC) was formed to regulate interstate communications (at that time, essentially AT&T).

In the 1980s AT&T was obliged by the US government to divest itself of its LECs, retaining its long-haul and manufacturing businesses. The 22 'Baby Bell' LECs quickly formed themselves up into seven *regional Bell operating companies* (RBOCs). A legal instrument called the 'Modified Final Judgement' gave the RBOCs and ILECs exclusive rights to provide local services,[7] while encouraging competitive long-distance *inter-exchange carriers* (IECs). During the 1990s the RBOCs merged until there were just four of them, and then the 1996 Telecommunications Act allowed *competitive LECs* (CLECs) once more.

Who's who in the marketplace

Figure 1.1 shows the main groupings of participants in the telecoms market. There are end customers, who want telecommunications services, and who buy such services from telcos. The telcos buy their network equipment from communications equipment manufacturers, who also sell terminal equipment direct to the telcos' customers. Various authorities regulate the doings of the telcos. The telcos, end customers and equipment manufacturers are all parasitized by IT companies. Lastly, the telcos, equipment manufacturers and IT companies organize themselves into all sorts of standards bodies and other gangs, typically to strengthen their trading

[7] This was in the belief that a monopolistic position was necessary to justify the investment required to meet universal service obligations in the local area.

positions through promoting common technical standards development.

Customers

There are of the order of 1 billion[8] telephone lines installed worldwide. The density of telecoms network access points is called teledensity.

$$\text{Teledensity} = \frac{\text{number of network access points (telephone lines, mobile phones, etc.)}}{\text{size of population}}$$

Teledensity averages about 15 per cent globally but is not by any means uniform across the world. In parts of North America, Europe and many of the Pacific Rim countries, teledensity exceeds 40 per cent. In other parts of the world, teledensity is 1

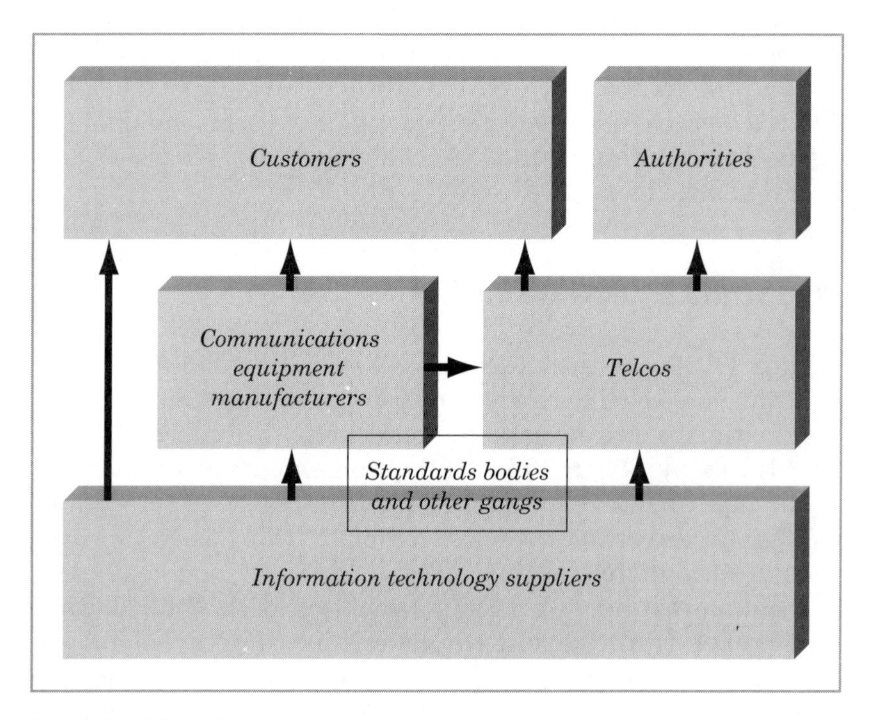

Figure 1.1 Who's who

[8] Throughout this book, 'billion' means 1000 million, not the older meaning of the word.

per cent or lower. Most current telecommunications revenues come from a small fraction of the human race, covering a small fraction of the globe.

Authorities and gangs

Telecoms companies are controlled by two quite different kinds of external organization. On the one hand, there are a small number of 'authorities' set in place by governments to control the market, with real legal power in their territory. On the other hand, there are a large number of standards-setting bodies, mostly without legal powers. These nonetheless can exert immense power when they get the market behind them. Telecoms systems have to interwork with each other to be of any use, and if it appears that most of the market is going to follow the lead of a particular standards body, that body acquires great influence over the market.

Telecoms systems have to interwork with each other to be of any use.

Global

At the global level, probably the most important controlling body, from the engineer's perspective, is the International Telecommunications Union (ITU). The ITU (formerly known as the CCITT) is an arm of the United Nations, and includes representatives of national governments, telcos and equipment manufacturers. Its primary job is to promote harmonization of technical interface standards between network operators, so that the world's networks can interoperate. A secondary result of this is that the ITU's standards are widely implemented by the manufacturers, and widely used within networks as well as at their edges.

The ITU's technical work is split between two groups: the ITU-T (Telecommunications) and the ITU-R (Radiocommunications, formerly the CCIR). The ITU-T calls its standards 'recommendations' and divides them into series relating to common themes. ITU-T recommendations are identified by their series letter, followed by a number. So M.3100, X.25 and T.37 are all examples of the ITU-T's output. The ITU-R, as well as producing technical standards relating to radio interfaces, runs WARC, the World Administrative Radio Conference, which meets every five years to establish *radio frequency* (RF) spectrum allocations worldwide.

Also at the global level is ISO, the International Organization for Standardization.[9] ISO has a very wide scope, and impinges on the telecommunications market mostly through its computing standards such as the *open systems interconnection* (OSI) data communications framework.

The global Internet is supervised by the Internet Society (ISOC), which includes representatives from many parts of the Internet community. The ISOC works through a number of subsidiary bodies, including:

- the Internet Engineering Task Force (IETF), which supports the technical development of the Internet by publishing technical standards, demurely called *requests for comment* (RFCs);
- the Internet Architecture Board (IAB), which provides technical guidance and strategic direction to the IETF;
- the Internet Engineering Steering Group (IESG), which manages the standards-setting activities of the IETF.

There are a number of other global organizations producing technical standards for the Internet. Standards for *hypertext markup language* (HTML) and other WWW technologies are defined by the WWW Consortium (W3C).

Blocks of Internet addresses are assigned to subnetworks under the authority of the Internet Assigned Numbers Authority (IANA), and the Internet Network Information Centre (InterNIC) manages the allocation of domain names through various lower level agencies, as described in Chapter 8.

Regional

To begin with, telecoms networks developed in isolated fragments, in many countries. Interconnection of networks between countries or continents came later. Therefore, it is not surprising that a number of different network technologies developed.

For social and political reasons, it was easier to address the harmonization of network technologies at a regional level than the global level. This led to a large number of regional

[9] That's right, the words do not match the abbreviation.

standards bodies being formed. Now that telecommunications is a global business, regional standard-setting has less value, and so we may see a tendency for the regional bodies to globalize or disappear. For now, though, there are a lot of them. Here are some of the better known ones. Others are mentioned where relevant in the following chapters.

In North America, ANSI, the American National Standards Institute, represents (among others) US computer manufacturers and users, and is the US member of ISO. ACTA, America's Carriers Telecommunications Association, represents the interests of more than 100 small IECs. The National Bureau of Standards (NBS) publishes the Federal Information Processing Standards (FIPS), which the US government procurement agencies use.

The telecoms engineer often comes across 'Bellcore' standards. The Bell COmpanies Research and Engineering organization was a part of AT&T until its restructuring in 1995. Bellcore produced the standards that all the Bell companies used and demanded of their manufacturers. Bellcore standards therefore commanded a huge market and were very influential, even beyond the US. Following its separation from AT&T in 1995, Bellcore continues to operate, but under the name of Telecordia.

European regional standards are produced by organizations including CEPT, the Conference Européen Des Postes et Telecommunication, ECMA, the European Computer Manufacturers' Association, and ETSI, the European Telecommunications Standards Institute. ETSI in particular publishes most of the wireless telecommunications standards for Europe, and ensures co-operation with the global standards bodies. ETSI standards are called European Telecommunication Standards (ETS).

State governments

At the level of individual states, the focus is usually on market regulation rather than technical standard-setting. In the UK, the government has delegated the management of competition, pricing and licensing to the Office of Telecommunications (Oftel), while a separate body, the British Board of Approvals for Telecommunications (BABT), is responsible for approving equipment for connection to public networks.

Other gangs and consortia

Established standards bodies are not always as focused or as quick as they might be. Therefore, there is a tendency for any new communications technology to give rise to its own association of network operators and manufacturers, with the purpose of quickly defining technology standards so that the technology can be brought to market. Examples of such associations are the GSM Memorandum of Understanding organization, the TeleManagement Forum, the ATM Forum, the IN Forum, Bluetooth and Radiccio. Generally these associations do very well for a few years, then begin to slow up. At this point they either hand over their work to a regular standards body, or lapse into exactly the sort of wooliness and lack of productivity that they set out to circumvent.

Telcos

Telcos are of several flavours, and they co-operate and compete in complex patterns. Figure 1.2 illustrates the main flavours and shows how they relate to one another.

Circuit mode network operators

Until the last quarter of the 20th century, all the telcos operated predominantly circuit mode networks, and many of them provided both access networks and long-haul connection, as well as doing their own sales direct to their customers. There are still a number of PTTs and ex-PTTs which cover the whole market in this way, but it is becoming less common. Since the progress

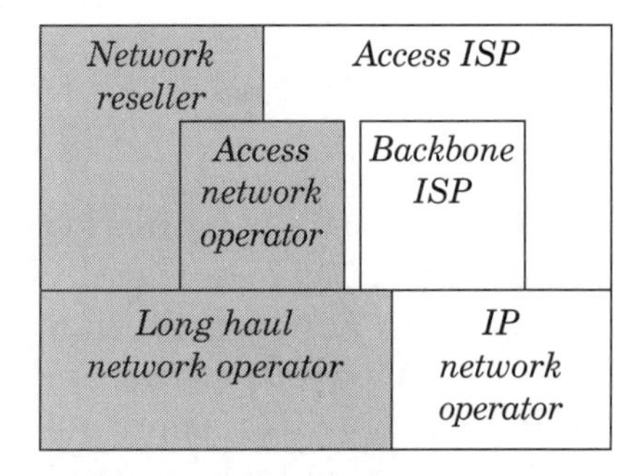

Figure 1.2 Telco flavours and relationships

of deregulation, circuit mode telcos have in many cases chosen to address only selected parts of the market. So there are national and global long haul specialists, which sell interconnect services to access telcos, network resellers or direct to corporate customers. There are access network operators, which specialize in covering a geographical territory (for example, the LECs in the USA, or *community antenna* TV – CATV – franchise operators) or in a particular access technology such as GSM (*global system for mobile*) or LMDS (*local multipoint distribution service*). But many of the larger access network operators find it cheaper to have their own long-haul network (where local regulation allows) than to buy connectivity from other operators, so they end up with their own national networks.

In recent years, a combination of deregulation and advances in IT have allowed a new breed of telco, the 'indirect telco' to develop. As Figure 1.3 shows, it is possible to be a telco without having the full panoply of access network, trunk network, switches, billing, service and sales organizations.

First, what can you do without a network? For example, *international simple resale* (ISR) operators buy time on international links, in bulk, from long-haul operators, and then resell it, using their own switches to direct the traffic.

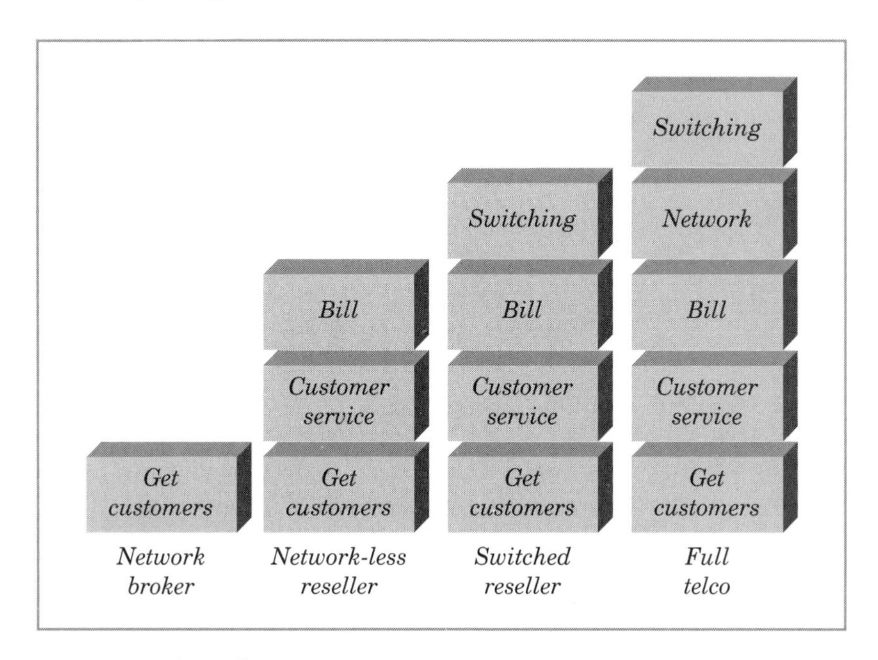

Figure 1.3 Indirect telcos

Callback operators offer cheap international calls without even bulk buying. They take advantage of differentials in international call rates. For example, it may be cheaper to call country A from country B than to call B from A. If a customer in A wants to call somebody in B, they place a call to a callback operator in A, specifying the number that they want to call. The callback operator arranges for its switch in country B to set up the call from that end, and charges the customer the B-to-A rate, plus a premium. Personal number operators have switches which just translate between a customer's 'personal' number and the number that they happen to be available on at the time, and then route the call back into the network.

It isn't even necessary to have switches to be a network service provider. Many mobile network operators, for example, choose to outsource their sales, service and billing operations to airtime resellers, from whom the public buys the network service.

Even this level of complexity is not strictly necessary. A number of network broker operations have been set up to facilitate the resale of network capacity between operators. They have no high-volume end-user billing or customer care to do; they just need to attract customers, provide a simple commercial service, and take their mark-up.

Packet mode network operators

By far the most significant packet mode network is the Internet.

By far the most significant packet mode network is the Internet.[10] In contrast with the carefully planned and slowly developed networks of the circuit mode telcos, the Internet is a chaotic and scarcely mapped patchwork of some hundreds of thousands of subnetworks, evolving at great speed without a clear high-level plan.

The Internet originated as an overlay on the networks of the circuit mode telcos. 'Backbone' *Internet service providers* (ISPs) buy long-haul links (as E-1/T-1 or above) from circuit mode telcos, and operate *internetwork protocol* (IP) over them. They provide high-capacity switching and route management, and

[10] 'Internet', with a capital, is widely used to refer to the global system of interconnected packet mode networks; 'internet', without a capital, is used to refer to any network or technology based on the *internetwork protocol* (IP), whether or not it is a part of the Internet.

sell IP capacity to corporate customers or to access ISPs. Access ISPs buy IP capacity, add customer-facing features such as portals, customer care and dial-up access, and sell to end users. More recently a number of IP network operators have appeared, with their own global fibre networks, running nothing but IP, and selling to the ISPs.

Common to all Internet operators is the concept of charging by connection time (or possibly by volume of data transferred), but not by distance. This has contributed to the downward trend in the circuit mode telcos' prices for long-distance transmission. Already a number of ISPs are providing a free service, even to the point of supplying free PCs for users. It is widely expected that the new commercial structures that the Internet has brought will lead to network connectivity becoming free to users and paid for indirectly by the providers of advertising, home shopping or other commercial content.

Equipment suppliers

Until the second half of the 20th century, the telcos generally made their own equipment, or contracted out the manufacturing of their own designs. The first generation of telecoms equipment manufacturers grew up around the development of telephony switches. Because these were large and complex machines, their development and manufacture required immense resources, and so the first-generation equipment manufacturers generally became very large organizations.

In the 1990s a second generation of telecoms equipment manufacturer appeared. Data communications manufacturers, which had grown up in the 1980s as vendors of enterprise networking kit, suddenly found their wares in demand for the huge Internet market. Some of these manufacturers were snapped up by the first generation, to maintain its credibility in the IP world, but others survived to dominate the telecoms equipment market by the turn of the century.

IT companies in telecoms

IT companies operating in the telecoms arena are in most respects similar to, if not the same as, those operating in other sectors. There is the usual mix of computer manufacturers, software vendors, systems integrators, consultancies and facilities management companies. Some of them offer products or services which are genuinely telecoms-specific, but most of them offer generic IT, with a little adaption to the interests, language and business processes of the telcos.

Market conditions and trends

Scale of the market

The global telecoms market is very large. In 1999, global service revenues in the telecoms market were around $1000 billion, which is comparable to the whole gross domestic product of the UK. Telcos vary widely in size. The largest ones turn over $80 billion or more, while local network operators may turn over as little as $100 million, and ISPs and resellers still less. The largest telecoms equipment vendors turn over $40 billion or more, which means that they are often as large as, or larger than, the telcos they are selling to.

Growth

Overall, telecoms service revenues are growing a few per cent faster than the world economy. This growth is, however, not evenly distributed across service types. Fixed-line and voice-traffic revenues are growing modestly, only a little more than the economy. Most of the growth is in mobile service revenues[11] and data communications services.[12] Most authorities agree that, as shown in Figure 1.4, the market for data services will at some point in the first decade of the 21st century match and then rapidly dwarf the voice services market.[13]

Growth is not uniform across the globe. Uptake of new services continues to be strong in the existing markets of North America, Europe and the Pacific Rim. However, the highest rates of investment in telecommunications infrastructure area are in regions which entered the 21st century with growing economies and underdeveloped communications infrastructure, notably China, the countries of the former Eastern Bloc, and South America.

[11] Variously estimated at between 25 per cent and 65 per cent per annum.

[12] Estimated wildly at anything from 30 per cent to 300 per cent per annum.

[13] That is, the market as measured in money. If you measure the volume of traffic carried, data has already overtaken voice, but it generates less revenue per volume.

Connectivity no longer the business

Many telcos have realized that owning and operating a network is not in itself a great money-spinner. The market price for simple connectivity has been falling since the 1980s, driven down by increased competition and by improvements in transmission technology. Even the dramatic rise in data traffic demand fails to make connectivity a good business to be in, because the unit price of data transmission is falling about as fast as demand is rising. Connectivity is in many cases being given free to the end user, as an aid to delivering higher-value services such as electronic commerce. It is therefore important for telcos to offer a service that is more than just the network connectivity. Telcos are refocusing on what they now see as their core business of building customer services out of networks and other elements. The networks are becoming a commoditized, low-value burden which the telcos are happy to outsource.

Many telcos have realized that owning and operating a network is not in itself a great money-spinner.

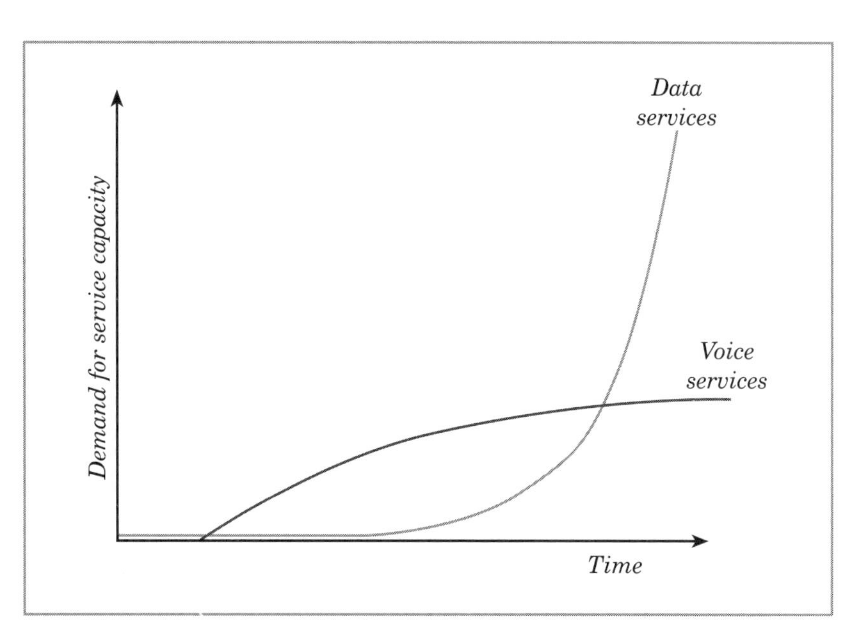

Figure 1.4 Voice and data service demand

Verticalization

Telcos, realizing that content is more profitable than connectivity, are scrambling to establish themselves as content providers. Sometimes this is by direct provision of content (for example, special Web portals). More often it is through purchasing content from third parties and retailing it under the telco's own brand. However, as internet technology allows the user, in many cases, to just use the telco's network for connectivity, and seek the best content elsewhere, the success of the telcos' approach is not assured.

Convergence

Between 1965 and 2000, a number of technologies and businesses had shown signs of gradual convergence with telecommunications. Around the turn of the century, advances in transmission and computing technology accelerated the process markedly. Video broadcasting can share the same transmission line as voice telephony and offer interactive features like the WWW. Voice telephony is no longer separate from data networking, but can use the same network infrastructure. Mobile radio access technologies can offer the high-speed data access previously restricted to fixed connections. The Internet and the telcos' networks are becoming the same thing.

From an engineering point of view, this convergence is probably beneficial because it makes technologies from other areas available for use in telecoms systems. From a business point of view, it's very confusing; many companies scarcely know what business they're in any more, and are being forced to try more and more risky ventures in chasing after the market.

Supply relationships

System procurements in telcos

The relationship between the telcos and their equipment vendors has recently developed rather quickly. Until the wave of privatization and deregulation in 1980–2000, a telco would commonly have a single preferred supplier, or perhaps two, for network equipment. Then, under the competitive pressure of the new markets, many telcos started to use multiple suppliers and wider competitive tendering. Having multiple suppliers proved to be a nightmare because it led to heterogeneous networks and thence to very high maintenance and support costs, and because of the high cost of managing multiple business relationships.

Therefore, many telcos tried dual-sourcing instead. The logic was that two competitive suppliers would be enough to keep prices down, without introducing excessive management or support costs. Dual-sourcing worked up to a point, but was imperfect because of the difficulty of finding two suppliers with exactly matched product sets. In real life, say, supplier A's switch would have some different features from supplier B's, so the choice of supplier would rest on the detail of the technical requirements and existing apparatus, rather than on price.

As it began to become apparent that owning and running a network was not necessarily a particularly profitable game, telcos began to look for ways to get their equipment suppliers to take over the burden. So, in some new telco start-ups of the late 1990s, the telco has chosen a single supplier, to supply, install and operate the network. To keep the supplier motivated, the deals involved have been on the basis of the supplier having to share risk with the telco, and not be paid until the network is working, or even not until the telco makes a profit. After all, as the equipment suppliers are often bigger than the telcos, they are just as able to take a few hundred million dollars of risk.

Telco mind sets

Telcos have their own special characteristics, which make them interesting, to say the least, to deal with. They are usually acutely technically aware, and talk in the sort of techno-speak and acronyms that this book introduces. They usually have huge amounts of money to spend. Telco systems procurements worth hundreds of millions of dollars are commonplace, and multi-billion-dollar deals are not unusual. For the software engineer this means that, whatever the software costs, it is unlikely to be a very large item on a telco's shopping list.

Because of their size and technicality, telcos are also often arrogant. Many of them have their own research and development organizations, and their own software engineering departments. They often find it hard to believe that anybody in the whole wide world could possibly do telecoms software engineering as well as *they* can. However, interestingly, there is often a divide within a telco, between the IT department and the people who run the network. The latter group is often cynical about the former, and much more ready to seek outside help if they can get away with it.

What telcos want

Telcos generally seem to want the same things. They want to gain new customers. They want to reduce churn.[14] They want to introduce new services before their competitors do. A typical national telco may be developing as many as a hundred new service offerings at a time. The time frame for inventing and introducing new services has been shortening for the past 15 years or so. Now a service may have to be in operation within just a few weeks, or even days, of its conception. There is therefore no longer much of a place for extended software development projects. A software development for a telco has to show a return within a few months at the very most. The telcos' horizons are short; five-year plans are meaningless because no telco today knows whether it will be in business in five years' time. The inefficiencies of taking a short-term view are well understood, but have to be accepted because of the velocity of the market.

Telcos demand not only that software development projects should be done in breathtakingly short time, but also that they should deliver software which is as reliable as the rest of their network, and that every project should come in on time. This sort of pressure is found in some other industries now, but the telcos have it more than most.

Figure 1.5 shows a typical maturation process for a telco. When a telco starts up, it has to satisfy its investors by building its network (and/or other technology) and getting a satisfactory number of customers. Typically, a substantial customer base has to be acquired within the first year of operation if the investors are to be kept happy. The telco's focus is therefore very clearly on the present. All kinds of opportunistic expedients are adopted, if only they help to get the operation going

[14] 'Churn' means a customer leaving their current telco service provider and either taking their custom to a competitor or giving up the service altogether.

and the customers signed. Details such as network quality, customer care, or efficiency of operational processes are secondary. Systems and software have to be immediately available off the shelf; there is no time for custom development, and no troublesome legacy of existing systems to integrate new systems with.

As a telco matures (if it survives that long), its attitudes change. Its investors start to expect it to make a profit, so it has to use its resources more efficiently. Customers begin to churn, so the telco has to sort out its network and service quality. New services have to be introduced, to keep up with the market, but to control costs, the telco has to re-use the existing network infrastructure. The opportunities for interesting custom software engineering begin!

Customers begin to churn, so the telco has to sort out its network and service quality.

Software vendors' supply model

Software products

Telcos have had their fair share of software procurement disasters. A fairly common theme of these disasters has been the involvement of software services vendors, developing bespoke or semi-custom software. Therefore, there has been a trend for

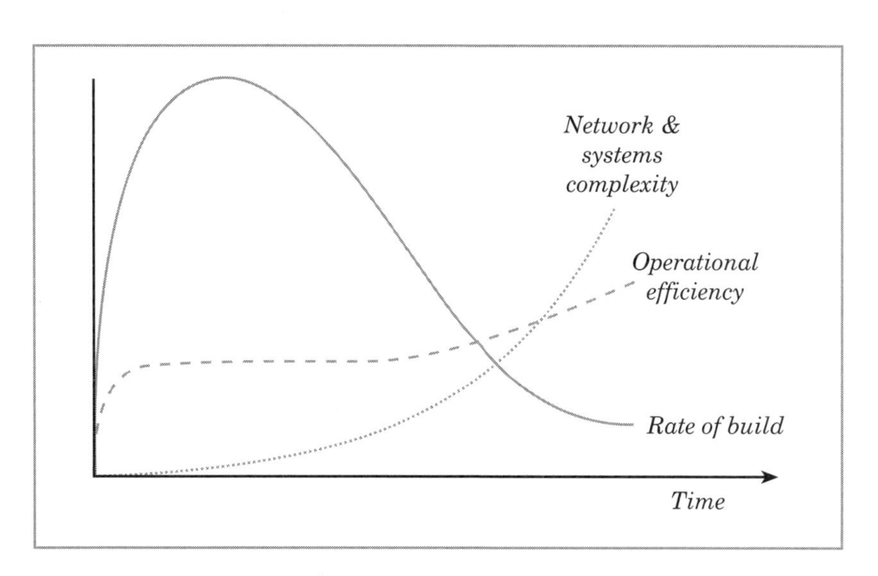

Figure 1.5 Telco maturity

telcos to seek to minimize their risk exposure by avoiding custom engineering, and buying only off-the-shelf products.

This has had two consequences. Firstly, it has become essential for any firm that wants to make big money in the telecoms sector to have a product to offer. For this reason, most of the systems integrator companies keep on either buying up, or making distribution deals with, developers of interesting telecoms software packages. Also, many software services organizations hold on to the *intellectual property rights* (IPR) for the system that they developed for telco A, so that they can label it as a product to sell to telcos B, C and D. From a software engineering point of view, this can be frustrating, because a piece of software designed as a one-off system is often awkward to convert into a multi-copy product, and so the product quality and lifetime can be compromised.

Interestingly, it doesn't seem to matter that the products are often Trojan horses, sneakily introducing lots of customization work. At the procurement stage, the telcos are often strangely willing to forget their past experience, and allow themselves to believe that the ready-made product will do exactly what they want, without adaption.

Software services

Software services suppliers have had to adapt to trends in the telcos' attitudes, in particular towards time scale and risk. Back in the 1980s, many telcos tried to control the risk in software development by insisting on a full and detailed specification of technical requirements, followed by a fixed-price contract for the implementation. This thorough and superficially wise approach tended to lead to inflation of requirements (to make sure that nothing was missed out before the implementation contract was let) and consequently to long development projects, often of two years or more. The trouble was that this attempt to remove one kind of risk just led to another. Over such long project lead times, the likelihood of the telco's requirements changing was immense, and so the telco faced a very high risk that the system it ordered would no longer be suitable by the time it was delivered.

A gradual realization of this problem, coupled with the acceleration in service development, has led to many telcos moving away from the big-bang fixed-price approach. Now many telcos let out development work in small lumps, under flexible *time and materials contracts*. This approach gives the telcos the flexibility they need, while not exposing them to much risk; if the software development goes wrong, the loss is relatively small, and the telco can try another supplier next time.

Software process qualification

Most mature telcos have come to take seriously the issues of software quality management. Several of the largest ones are actively involved in international software process quality assessment programmes. Two examples of these are:

- the Software Engineering Institute's (SEI) *capability maturity model* (CMM), which scores an organization's ability to monitor its software engineering processes, to achieve repeatable results, and to continually improve itself;
- ISO 9001, a standard for process definition across a wide range of industries. Of particular interest is ISO 9001 T, an emerging standard for telecoms systems in particular.

Further reading

For a more detailed telling of the American story, see Cole (1999). Readers interested in the history of the market should look at the early chapters of Goralski (1995). For current market information, you will have to consult trade journals, of which there are plenty; at the time of writing, *Tele.com* and *Network Fusion* were two of the more market-aware ones.

2 Network architecture concepts

Here are some ways of describing, partitioning or classifying networks. They are largely independent of the network technology, and so their usefulness persists despite technical change.

Multilayered model

Figure 2.1 presents a way of looking at networks as several superimposed layers. At the bottom, the only real physical network is the transmission network. This comprises the physical transmission media (say fibre, or wireless cells), the media access equipment (line termination devices, for example) and multiplexing equipment. The switching nodes at the interstices of the transmission network, and the connectivity which the

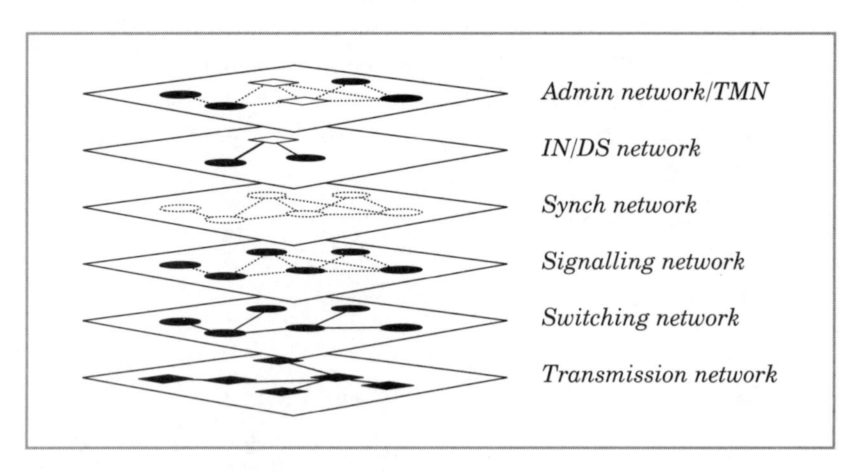

Figure 2.1 Layered (virtual) network model

transmission network gives them, can be viewed as defining a distinct *switching network*. Because some transmission systems use complex network structures to achieve superficially simple connectivity, the switching network often has a topology that is different from, and simpler than, the transmission network.

At the next level up, the *signalling network* is a view of the ways in which the switching nodes choose to signal to each other. As they generally will not use all of the network connectivity for signalling, the signalling network has its own distinct shape.

Shown above the signalling network, but really at a similar level in the network structure, is the *synchronisation network*, a system of transmission links for clock distribution across the switching and transmission networks.

Superimposed on the switching network is also often an IN (*intelligent networks*) network, sometimes called the derived *services network*, a system of interconnected computers which use the facilities of the switching network to offer advanced facilities such as number portability.

Lastly, all these systems and networks need to be supervised, and as telecoms network are geographically widespread things, there has to be some kind of *telecommunications management network* (TMN) or to enable unified surveillance and control of the whole network.

Circuit mode and packet mode networks

Early, *circuit mode*, network technologies assigned a full complement of network resources to a circuit for its whole lifetime, whether it was carrying user traffic or not. For example, when a telephone call is made, the network sets up a bi-directional digital circuit to carry the voice signals. All the network equipment that the circuit uses is tied up until the call ends, even if there are long silences in the conversation. Clearly, circuit mode networks can be very wasteful of resource, particularly if their users are teenagers.

More recently, *packet switching* has been introduced. In a packet mode network, the user data (for example, the digitally encoded speech) is sent across the network in small chunks called packets. While there is no data to send, no packets are sent, and the network resources can be used for other things. Many packet mode technologies support the use of *virtual circuits*. A virtual circuit is one where the network has committed to provide communications between two end points, but assigns resources to it only when there is traffic to carry. There are *permanent virtual* circuits (PVCs), where the users can send traffic without any preliminaries, and *switched virtual circuits* (SVCs), where the users request a circuit, use it for a while, and then relinquish it. Both PVCs and SVCs are classed as *connective* because the network recognizes a connection between the end points.

Alternatively, a packet mode network can offer a *connectionless* service, where each message (or *datagram*) has to be marked with the address of its destination, and network resources are tied up only when a message is being transmitted. Transmission systems, both circuit mode and packet mode, can also be characterized by their number of receiving parties. Ordinary one-to-one communication is called *monocast* or *point to point*. Transmission to a defined set of destinations, as when a fire chief gives a direction to his whole team, is called *narrowcast* or *multicast*. *Broadcast* transmissions are those which are directed to any suitable receiver.

Packet networks have made an enormous impact on network design and economics.

The division between circuit mode and packet mode technologies is very significant. Packet networks have recently made an enormous impact on network design and economics.

Generic network architecture

The network architecture shown in Figure 2.2 originated from the Public Switched Telephony Network (PSTN), the worldwide fixed-terminal voice network, which dominated the telecoms business until the 1980s. Despite the enormous changes in network principles since then, the architecture is still very widely applicable, and provides some of the most fundamental terms used in telecommunications.

Large numbers of *user terminals* are connected to a telco's network via *network termination equipment* (NTE), for example the box on the wall where you plug your telephone in. Sometimes the telco provides equipment to connect to the NTE within your premises, for example a PABX (*private automatic branch exchange*). Such telco equipment lodged within your premises is called *customer premises equipment* (CPE). The

NTE connects to an *access network*, which reaches from the user's premises to the telco's nearest switching node. The access network typically covers a distance of three or four kilometres; for this reason, it is often loosely called the *last mile*. It is also called the *local loop* (an allusion to copper loop telephony transmission media) or *the local distribution network*.

The telco's network contains a number of switches, also called exchanges or central offices. Some of them are connected to the access network, while others handle only intra-network (transit) traffic. The switches are connected across the telco's territory via a trunk (or long-haul, or backbone) transmission network.

In order to provide connectivity between users on each other's networks, the telcos have to have interconnections between their trunk. These are called *points of presence* (POP). At a POP, the telco at the far side of the connection is often called an *other licensed operator* (OLO).

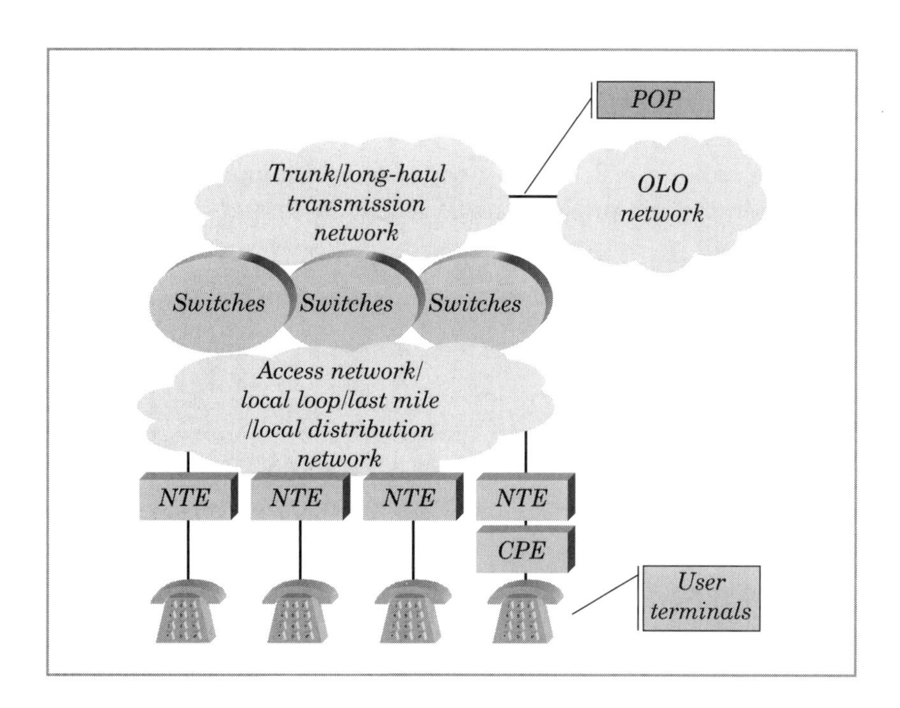

Figure 2.2 Generic network architecture

Reference model for transmission systems in the access network

A number of reference models have been produced (by the ITU-T, ETSI, ATM Forum, DAVIC and others) with the aim of defining a standard terminology so that when we talk about the various parts of an access network, we can use terms which are the same regardless of the transmission technology. As all these reference models are different, the enterprise has not been an unmitigated success. However, there are a number of shared features that appear in most of the models. These are presented crudely in Figure 2.3.

Working from right to left, Figure 2.3 shows:

- ET (*exchange termination*), the switching node(s), in a local exchange building to which the access network connects;
- LT (*line termination*), the item of equipment (for example, a modem) at which the access network terminates in the exchange building;

- the V reference point, also called the *service node interface* (SNI) or *access network interface* (ANI), where the ET connects to the LT;
- NT1 (or plain NT), the network termination where the access network presents itself in the customer's premises;
- the U reference point, which refers to the arrangements for the access network between the LT and the NT1;
- NT2, the item of customer's equipment which connects directly to the NT1;
- the T reference point, also called the *user/network interface* (UNI), at the interface between NT1 and NT2;

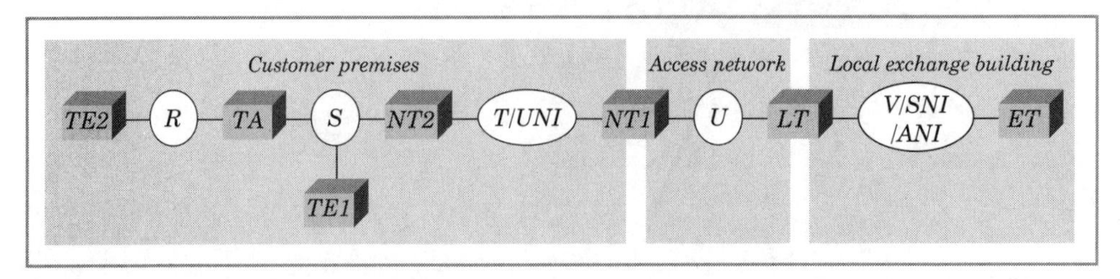

Figure 2.3 Composite reference model for the access network

- TE (*terminal equipment*), which is what the customer connects to the NT1; sometimes TE is subdivided into TE1 (standard, plug-and-go TE) and TE2 (non-standard TE, which needs some kind of *terminal adapter* (TA) to make it work with an NT2);
- the S reference point, between the TE1 and the NT1;
- the R reference point, between the TA and the TE2.

These terms are used ad nauseam in the standards documents, and even in this book I do not escape using them now and then. In particular, the UNI and the V reference points are handy common reference terms.

Basics of transmission

While the reader may not be expecting to work in the detailed design of transmission systems, it is useful to understand the basics of analogue and digital transmission.

In telecoms, 'transport' is nothing like the same thing as the 'transport layer' of the OSI seven-layer model. It just means getting a stream of bits from one point to another, without any concern for their significance, grouping or formatting, and often without any very special care about their integrity. Transmission systems are the technologies that transport these bits.

Analogue and digital

An *analogue* signal is one which can vary, with time, over some continuous range of values, and where the transmission system is concerned to communicate an exact replica of it, or as near as it can get. Figure 2.4 illustrates an analogue signal generated by a musical instrument.

A *digital* signal, rather than having a continuous range of values, is composed out of a discrete set of recognized values or symbols, as in the letters of this sentence. Considerable variations in the quantities transmitted can be made, **like this text** so long as the receiver can tell which symbols are intended.

The usual measure for the strength of a signal is not its amplitude but its power.

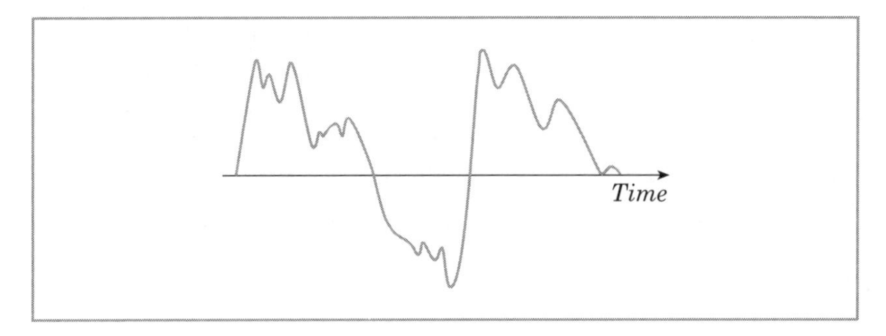

Figure 2.4 An analogue signal

Signal parameters

A signal's *amplitude* is the measure of how big a range of values it varies over. However, the usual measure for the strength of a signal is not its amplitude but its *power*, which varies more or less as the square of its amplitude. Signal power levels can be compared with each other in *decibels* (dB). The decibel is a relative measure; you can say that the power ratio between signal A and signal B is –6dB, but you can't use dB to say in absolute terms how powerful a signal is. The most common power level unit in electrical signalling is the dBm, which says how powerful the signal is, compared with one milliwatt.

If a signal has a repeating structure, like the one shown in Figure 2.5, its *wavelength* is the measure of the distance between repeats, and its *frequency* is the number of repeats per second. As the signal is usually travelling along in space, wavelength is often specified in metres, millimetres, etc. Frequency is properly specified in s^{-1}, but more casually as Hertz (Hz), or kilohertz (kHz), megahertz (MHz) or gigahertz (GHz).

Also, it is possible to have several signals, which are in most respects the same, except for a time delay between them, as Figure 2.6 illustrates. The difference in time between the signals is expressed as a fraction of the wave period, and called either the *phase difference*, or more commonly, the *phase angle*.

A strictly repetitive signal like the one in Figure 2.5 is easy to analyse mathematically, but it conveys very little information. Once you've seen one wavelenth of it, you've seen them all. So analogue signals that convey useful information, for example the sound of speech, are not repetitive. However, signal engineers can use tricks called the Fourier transform, Laplace transform and Z-transform to analyse complex non-repetitive signals as if they were a collection of simple repetitive signals added together.

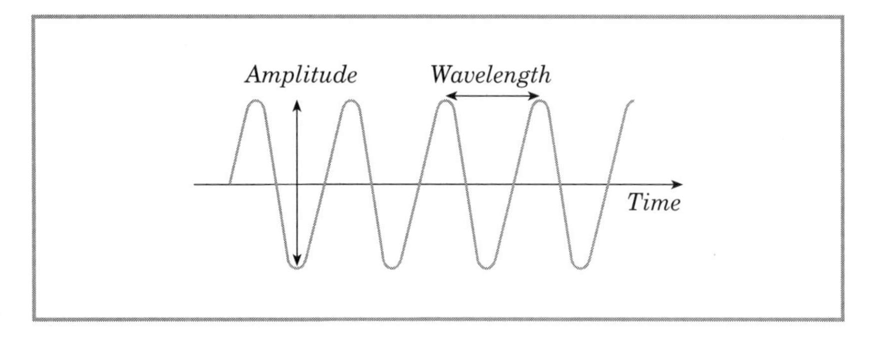

Figure 2.5 Signal parameters

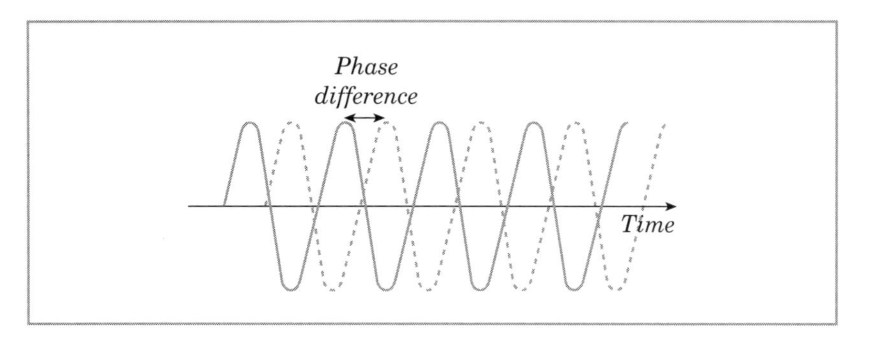

Figure 2.6 Signal phase difference

Transmission constraints

One of the challenges of transmission system design is that the signal that you get out at the receiving end of a link is never quite the same as the one that you put in at the transmitting end. There are various terms used for the things that go wrong along the way, most commonly:

- *attenuation* means a reduction in power, which typically depends on the length of the link; attenuation is measured in dB. The way the dB is defined (as the logarithm of a power ratio) means that you can work out the output power by simply subtracting the link attenuation from the input power;

- *a power budget* is a calculation to show the output power of a transmission system when all the sources of attenuation along the way are accounted for;
- *noise*, strictly speaking, means any signal which comes out that is not the 'wanted signal', the signal that you put in. So distortion, crosstalk and jitter, discussed below, are all kinds of noise. The power of all the noise sources on a signal path can be totted up, in a noise budget. Less strictly, noise is used to refer to extraneous and generally random signal elements, such as resistive noise;
- *distortion* covers systematic transformations of the signal along the way, for example the selective attenuation of some frequencies;
- *jitter* describes the situation where the signal that comes out is similar to the one put in, but where its timing relationship to the input signal is not constant; that is to say, the end-to-end time lag is varying;
- *crosstalk* is where the signal picks up traces of some other signal along the way, for example where two phone lines run in parallel and electromagnetic coupling causes the two signals to audibly interfere.

Electromagnetic spectrum

Almost all telecommunications signalling uses electromagnetic signals. Figure 2.7 shows how several signalling technologies fit within the electromagnetic spectrum. The range of frequencies that a signal uses is called its *bandwidth*. So for example, a mobile telephony channel might use a bandwidth of 200 000 Hz, between 900 000 000 000 Hz and 900 000 200 000 Hz.

Modulation methods

There are various ways of sending a digital message by using an analogue signal such as the voltage across two wires, or the brightness of light in an optical fibre.

In *baseband* signalling, the analogue signal is simply made to move between various agreed levels to represent the required symbols. Baseband signalling is however limited in value. Firstly, it allows only one signal at a time to be sent over a given channel; by contrast, carrier modulation schemes may allow several signals to be carried in parallel. Secondly, it can be unsuitable for many transmission media. An obvious example is radio signalling, where a high-frequency carrier has to be used to enable any transmission at all.

In *amplitude shift keying* (ASK), the amplitude of a constant-frequency carrier signal is made to vary. In ASK and all of the other keying systems discussed here, either two (binary keying) or several defined states may be used. Using a larger number of distinct states increases the number of bits carried per symbol, but makes the hardware design more demanding.

In *frequency shift keying* (FSK), the carrier amplitude is kept constant, but its frequency is made to vary. An everyday example of this is the horrible warble of a fax machine.

In *phase shift keying* (PSK), the amplitude and the frequency stay constant, and the signalling is done by varying the phase of the carrier. A specialized form of PSK, called *pulse position modulation* (PPM), is used in optical transmission systems.

These techniques can be used singly or in combination. One

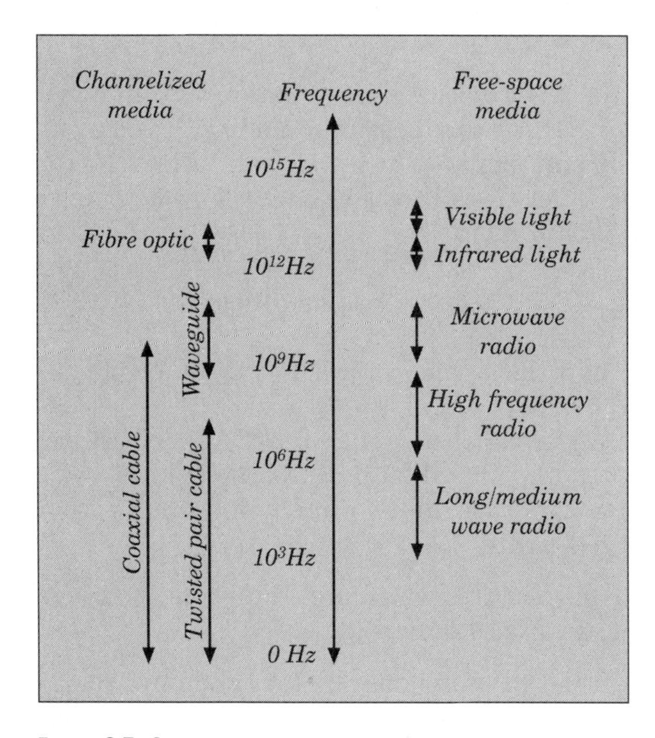

Figure 2.7 Spectrum

successful combined approach is *quadrature amplitude modulation* (QAM), where two copies of the carrier are transmitted, with a quadrature phase relationship (so that the peaks of one align with the zeroes of the other). Each carrier is separately amplitude-keyed, so if, say, four amplitude levels are used on each carrier, one of 16 possible symbols can be sent at a time.

With any form of keying, the choice of the number of distinct signal states to use is a trade-off between information content (n signal states encodes $\log_2(n)$ bits per signal element) and reliability (the more signal states, the harder it becomes to distinguish them in the presence of attenuation and noise). The number of distinct signal states (or 'symbols') per second is called the *symbol rate*, and is measured in *Baud* (symbols/second).

Bandwidth vs. Baud

The minimum bandwidth required to transmit B baud is 0.5 B, in theory. Practically, the bandwidth has to be at least equal to B.

The Shannon-Hartley capacity limit equation sets an upper bound on the error-free bit rate possible on a link: given available bandwidth B and signal-to-noise power ratio S/N, the maximum possible error-free bit rate or *channel capacity* (also called the *information rate*) is

$$B \log_2(S/N + 1) \textit{ bits per second } (\text{bps}).$$

We shall see later how this applies in practice to various transmission media.

Now that we have established the clear difference between bandwidth and bit rate, you will appreciate the silliness with which the industry confuses the two in the following rather loose, but much-used terms:

- *narrowband* means bit rates which are suitable for voice or low-speed data, say up to 128 kbps;
- *broadband* means bit rates of about 2 Mbps and upwards; the terms are so loosely used that their boundaries and the gap between them are always changing;

- *wideband* is nothing to do with bit rate; it is a reference to use of spectrum, in CDMA (*code division multiple access*).

Multiple access methods

Often a transmission link has enough capacity to carry several information channels simultaneously. For example, the radio bandwidth available to a mobile telephony base station is enough for many users' traffic. There are a variety of ways of ensuring that the channels do not interfere with each other. The broad classes of multiple access technology are as follows:

- the oldest method is *frequency division multiple access*, also called *frequency division multiplexing* (FDM). This works by allocating blocks of bandwidth, at different frequencies, to the different users or channels. So a cable, say, with a usable bandwidth of 100 MHz can carry 100 independent channels of 1 MHz each, or a bit less to allow for margins (*guard bands*) between the blocks of used bandwidth;
- with the arrival of integrated circuit technology, *time division multiple access* (TDMA) or *time division multiplexing* becomes possible. This means that each user uses all of the available bandwidth, but all the users take turns with it. Usually there is a repeating pattern of time slots agreed between the two ends of the link, and channels are assigned to one or more time slots, depending on their requirements. So, for example, up to eight GSM phones within a single cell can have the same bandwidth of 200 kHz allocated to them by agreeing with the control system which phone should use which of eight time slots.

Code division multiple access is a more recent technology which originated in battlefield communications systems but gained ground in civil telecommunications because of its tolerance to interference and its efficient use of bandwidth. The idea of CDMA is that all users transmit on all the available bandwidth. If there happen to be problems with transmission on some frequencies, their effect on the signal is minimized. The frequency spreading is done in one of two ways:

- in *frequency-hopped CDMA*, the signals are made to hop very rapidly around the available bandwidth, so that any signal problems caused by natural obstacles or interference from other users are very brief;

- in *direct sequence* or *spread-spectrum CDMA*, the user's digital signal is mixed (XOR-ed) with a spreading signal. The spreading signal is a binary pseudorandom sequence, with a clock rate (called its chipping rate, and defined in *chips per second* or cps) set so that it covers the whole available bandwidth. The combined signal similarly covers the whole bandwidth. The combined signal is modulated onto a carrier, and launched into the ether. All the users use the same frequency range, and they are all allowed to transmit at once. But the magic of CDMA is that each user's spreading signal is made to be different, and by demodulating using the right spreading signal, the receiver can eliminate the noise of other users' transmissions.

Torpedoes and Hollywood

The idea for CDMA was invented in the Second World War by the actress Hedy Lamarr. The polymath Lamarr approached the US military, offering CDMA as a solution to problems in the secure remote control of torpedoes. However, until the 1960s, CDMA was impractical because of its high demands on the signal-processing equipment, and until the 1980s it was classified military technology.

CDMA is complex, and calls for a high degree of co-ordination between the transmitting and receiving systems. However, its efficient use of bandwidth and transmitter power makes it attractive when these commodities are in short supply, for example in mobile phone systems. Another advantage of CDMA is that there is no hard upper limit to the number of users – the signal-to-noise ratio just gradually deteriorates as more users transmit.

Kinds of channel

So far, we have considered a simple one-way direct link between two systems. There are many ways of making things more complex.

The unit of end-to-end connectivity that a network provides to a user is generally called a *circuit*. A circuit can be implemented as a single link, or as a number of links connected end to end, at connection points called nodes. Other common names for a circuit are path (in the *plesiochronous digital hierarchy*) and *trail* (in the *synchronous digital hierarchy*).

A circuit can be *simplex*, which means that communication in only one direction is provided, as in the case of TV signal distribution, or it can be *full duplex*, in which both directions are supported simultaneously, as in normal telephony. It can be *half-duplex*, in which an alternation between the two directions is allowed, as in CB radio or PMR. Individual links can also be of any of these kinds. A full duplex circuit can, for example, be constructed as a concatenation of full duplex links, or as two parallel chains of simplex links. Where circuits are constructed out of simplex technologies, it is common for network designers to work in units of half-circuits.

A circuit can be semi-permanent, as in the case of a *leased line* or a *nailed-up* line. Circuits of this kind naturally have fixed end-points. Alternatively, a *switched circuit* is one that is set up and cleared down when its users request it, and where the end-points can vary, as in the PSTN.

Why digital transmission?

We have discussed ways of using analogue signals to carry digital messages. It may then seem perverse that when we need to convey an analogue signal such as voice, we generally do it by converting it to a digital representation for transmission. Why not transmit analogue signals as analogue signals, like Mr Bell did? There are in fact a lot of good reasons for using digital transmission.

- Any link will be subject to noise. With analogue transmission, that means that the signal is degraded with every link that it traverses. Before digital transmission, one had to expect that a long-distance or international call could not give anything like the speech quality of a local call, simply because of the number and length of links involved. In digital transmission, the keying scheme organizes the signal into a number of defined states. So long as there is not enough noise to confuse the receiver about which state is being sent, the received digital signal is *identical* to the one sent.

- This property of noise rejection permits regeneration of the signal as many times as necessary, without any degeneration. So very long circuits can be constructed out of a series of concatenated digital links, and the end-to-end quality can be largely independent of overall length. This is the magic that makes a call from across the world sound as clear as one from next door.
- The advent of digital networks was spurred by the invention of the transistor. When an analogue signal is carried, it usually has to be amplified at some point, to compensate for the attenuation of the transmission medium. To do this requires linear amplifiers; that is to say, amplifiers where the output is a magnified but faithfully reproduced version of the input. Before the transistor, amplification was done using valve technology, which is intrinsically fairly linear. Transistors, on the other hand, are intrinsically non-linear; a transistor's output signal is a very distorted image of its input. With transistors, it is cheaper to make a non-linear amplifier, and several times more costly to make a linear one. Non-linear amplification is no good for analogue signals, but more acceptable with a digital signal, where only a relatively few signal states need to be distinguished.

Using digital signals permits networks to be built with less concern for the intrinsic quality of the transmission links

- Because a digital signal can be recovered correctly, subject to the Shannon–Hartley limit, from a noisy channel, using digital signals permits networks to be built with less concern for the intrinsic quality of the transmission links, and so permits economies in link technology.
- Lastly, digital signals are convenient because they can be manipulated in computers.

Voice encoding

Codecs and pulse code modulation

Voice signals are analogue in nature, but most of the transmission and switching network is digital. The conversion is achieved by devices called codecs (a concatenation of *code* and *decode*), which are situated typically where the analogue access network reaches the first switching node, but which may alternatively be embedded in the user terminal if the access network is digital.

The first codecs implemented a process called *pulse code modulation*, invented in 1937 by Alec Reeves. A PCM encoder takes an analogue signal and samples it at regular intervals, then converts the analogue value of each sample into a digital, numeric form. A PCM decoder reverses the process. If the samples are close enough together in time, and if the values are recorded precisely enough, the original waveform can be reproduced to any arbitrary accuracy.

Logarithmic companding

In real life, there are limits to the rate and precision of sampling. The most common PCM schemes are called μ-law (used in Europe) and A-law (used in North America and Japan). They both implement a trick called logarithmic companding. μ-law, for example, samples the analogue signal at a rate of 8000 times per second, and quantizes each sample into an eight-bit word. Of the 256 possible values, 212 are actually used. These (and this is the clever bit) are not spaced evenly across the possible values of the analogue signal. They are set close together near the zero signal level, and further apart (logarithmically) at the extreme high and low values. This means that the quantization noise introduced by the sampling is roughly constant as a percentage of the signal value; to put it another way, quiet signals are sampled accurately, while the errors in sampling loud signals are small in comparison with the signal value.

μ-law PCM encodes the signal in (8 bits × 8 kHz) 64 kbps, whereas A-law encodes it (with 6 dB worse signal–noise ratio) in 56 kbps, in line with European and North American transmission rates respectively. Both schemes are defined in ITU-T recommendation G.711.

Low rate encoding (compression)

Since the cheap availability of computing and signal processing hardware, a number of systems for encoding acceptable-quality voice in lower data rates have been developed:

- *adaptive differential pulse code modulation* (ADPCM, specified in ITU-T recommendations G.726–727) encodes normal voice quality in 32 kbps;
- *linear predictive coding* (LPC) systems can encode speech adequately in as little as 9600/4800 bps. An example of the applications of LPC is in GSM, which uses it to encode voice in 13 kbps;
- voice-over IP applications use *linear prediction analysis-by-synthesis* (LPAS) encoding. The current favourite, G.723.1, uses LPAS to provide generally acceptable speech quality in 6.4 kbps, but with a processing delay of 60–70 milliseconds. G.729, while requiring more channel capacity, has a more acceptable delay of 20–30 ms, and may overtake G.723.1 as capacity becomes easier to command.

Further reading

Flood, J.E., ed. (1997) gives a clear and extensive presentation of network models generally, while architectural aspects of the access network are covered well in *BT Technology Journal* (1998). Access network reference models, and the status of standardization efforts relating to them, are discussed in mind-boggling complexity in ETSI EG 202 306 *Access Networks for Residential Customers*.

Shannon (1948) is worth knowing about because it presented, pretty much for the first time, the science of information theory, and very lucidly too.

Both Bateman (1998) and Halsall (1995) cover modulation and multiple access systems.

Voice encoding is covered briefly in Flood (1997) and Halsall (1995), and is reviewed in depth in Black (1999) and Douskalis (2000).

3 Fixed-node transmission technologies

Introduction

This chapter looks at the transmission technologies that work between nodes that are essentially fixed in location. Many of them can be called space-division or channelized technologies; that is to say, each transmission link has its own private physical space (for example, a wire or a glass fibre). Space division technologies are often loosely called 'wireline', although there may not be any metal involved.

Some other fixed-node transmission technologies are not channelized, and many nodes may be physically able to access each other's signals. These technologies share some of the access control problems of mobile terminal networks, but still escape the special difficulties of allowing nodes to roam from one part of the network to another. Those problems are addressed in Chapter 4.

Fixed-node transmission technologies are used both for trunk transmission, where the nodes involved are internal to the network, and for access network transmission, where many of the nodes are user terminals.

Copper

By far the predominant transmission medium worldwide is copper (Cu), whether measured by number of end points, distance covered, or value invested. This is because the first large-scale telephone networks used copper (some early efforts tried iron, which proved unsatisfactory). Nowadays copper is normally used only in the access network, and the original copper trunk networks have been superseded.

The essence of copper-based transmission is that an electrical circuit is established from end to end. Originally, these circuits carried an *alternating current* (AC) electrical signal, which was a direct analogue of the voice signal being carried. The circuit is typically a pair of insulated conductors, twisted together to help with noise rejection. Generally, the conductors are made of copper, but because that metal is costly, new installations now often use aluminium. Even then, the medium is generally spoken of as 'copper'.

Because the cost of running two conductors across the access network can be high, some network operators saved money by installing 'earth return' lines, where there is a single copper conductor, and the earth is used in place of the second conductor. However, such lines tend to be unreliable because of the difficulty of maintaining a low resistance connection to earth.

Another way for network operators to save money was to connect more than one subscriber across the same pair of wires, creating what is called a *party line*. A party line installation (which can still be found in some rural areas) can be recognized by the presence of a 'press to call' button on the telephone, which seizes the line (if it is not in use) and prompts the local exchange to generate the dial tone. Party lines are unsatisfactory in many ways, not least in that they cannot support both parties calling at once. They are becoming rarer.

Party lines are unsatisfactory in many ways, not least in that they cannot support both parties calling at once.

When the trunk networks used analogue transmission, they did this over copper pairs also. So that the signals could be amplified (in both directions) without interfering, the 'transmit' and 'receive' signals were split onto two separate pairs when they went from the local loop into the trunk network, and recombined at the other end. This was done by using transformers (called *hybrids*) which meant that although *direct current* (DC) signalling was possible across the local loop, there was no DC path from end to end across the whole network.

As a result, a normal (ITU-T Q.552) domestic telephone circuit offers a standard quality of service across the network which is characterized by a signal-to-noise level of 30dB and a bandwidth from 400 Hz to 4000 Hz. These frequency limits are not absolute cut-offs; rather, the frequency response rolls off steeply outside these limits. Within them, the end-to-end attenuation is specified to be –3 dB or better. While signalling outside these frequencies is possible, DC signalling is entirely impossible, so any PSTN end-to-end signalling system will have to work by modulation of a carrier.

The sordid mechanics of metal

An awful lot of what goes on within telcos is related to the mechanics of the access network, because that is the location of a lot of the trouble. Figure 3.1 introduces the major components of a copper access network.

The key elements illustrated by Figure 3.1 are:

- a *distribution frame* (DF) hierarchy. In this case it has just two levels: *concentration points* (CP), covering a few hundred subscribers, and a *main distribution frame* (MDF) adjacent to the local switch. There can be more. At each DF, it is possible to connect pairs from the multi-pair cable on the customer's side (the *distribution side* or D-side) to the generally fatter multi-pair cable on the other side (the *equipment side* or E-side). The connections are made using solderless insulation-displacement connectors. The DFs may be in underground chambers, or above ground in *street cabinets*;

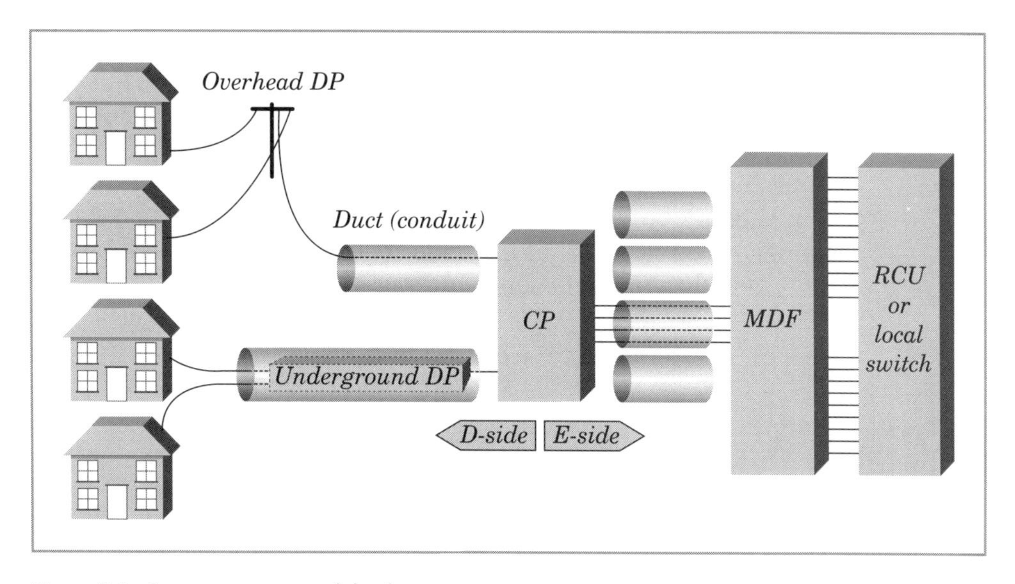

Figure 3.1 Copper access network hardware

- *ducts* (tubes) in underground trenches, carrying the cables part or all of the way from the customer to the local switch. Because of the risk of rainwater finding its way into the ducts and corroding the cables, the air in the ducts is pressurized from the exchange end. Not only does this reduce the chance of water seeping in, but it also means that pressure-monitoring equipment in the exchange building can detect duct damage by loss of pressure;
- *distribution points* (DPs), where the multi-pair cables are connected to single pairs going to the customers' premises;
- *overhead DPs*, or 'telegraph' poles. Historically, much of the access network was carried along chains of poles. However, improvements in trench-boring technology and increased environmental sensitivity have reduced overhead lines in many countries to just one 'distribution' pole, or a short series. Poles can be made solid, of wood or concrete, or they can be made hollow, which allows the distribution cables to be fed up the inside and out to the customers' premises with a minimum of overhead work.

The limits of the copper network

Globally, the copper access network is an immense fixed asset.[1] It often accounts for 50 per cent or more of a network operator's assets. As such, it is unlikely to be discarded for some time. However, it has a very serious limitation as a delivery medium for modern telecommunications. As we have seen, it offers a bandwidth of 3400 Hz and signal-to-noise ratio of 30 dB. Applying Shannon's law gives a theoretical maximum information rate of about 35 kbps across a PSTN. Getting data across a copper based network at the high speeds demanded, say, by digital video, is obviously something of a challenge.

Voice band modems

Modulator-demodulators (modems) use the end-to-end bandwidth offered by the PSTN to carry AC signals, which are modulated to carry digital information. The dominant ITU-T standard is currently V.90, which specifies very complex

[1] Allegedly there are about 43 million tons of copper built into access networks worldwide.

modulation schemes (including fancy stuff such as forward error correction and trellis coding) to offer data rates of 56 kbps.

This appears to contradict the result above, but in reality, Shannon's law is borne out. Fifty-six kbps modems are unreliable; on a very good connection they run at 56 kbps, but on a less favourable connection they fall back to 28.8 kbps or less.

Faster but shorter

It might seem as if Shannon's law and the 3400 Hz end-to-end bandwidth of the PSTN mean that data rates across a copper access network can never rise, but this is not so. The bandwidth limitations are end to end across the network, and are due largely to network elements that are beyond the access network. A copper *access network* is itself capable of 1 Mbps or more in theory. The following sections introduce a whole raft of technologies that are competing to make that capacity available to network customers.

ISDN digital subscriber line

Basic rate ISDN (BRI) technology, developed in the early 1980s, uses echo-cancellers and alternate-half-duplex methods to deliver 144 kbps over a single copper pair. This seemed fast at the time, and in fact the initial market uptake was very poor because of a shortage of really attractive high-data rate applications. The ISDN market didn't really take off until the mid-1990s, with the emergence of the Internet as a commercial tool. By then, newer and faster *digital subscriber line* (DSL) technologies were appearing. As a result, 144 kbps ISDN is now positioned as 'narrowband ISDN'.

The ISDN market didn't really take off until the mid-1990s, with the emergence of the Internet as a commercial tool.

The 144 kbps is presented at the S interface (via an eight-pin RJ45 connector) as two 64 kbps 'B' (bearer) channels and one 16 kbps 'D' (data) channel. These channels provided a single, common point of access for voice and data services, including:

- 64 kbps circuit switched data transmission;
- voice telephony, with faster call set-up and clearer audio quality than an analogue line offers;

- facsimile (supporting Group 4 fax machines);
- slow-scan videophone, CCTV and video conferencing;
- *teletex* (user-to-user message exchange);
- *videotex* (message-based access to a central data store, as in the UK 'Prestel' service);
- X.25 packet switching and ATM.

ITU-T recommendation I.411 defines a 'reference configuration model' for ISDN networks, as shown crudely in Figure 3.4. The signalling protocols which are defined for the S, T and V reference points are discussed in Chapter 6.

In some countries, higher rate ISDN services are available, under the general title of *primary rate ISDN* (PRI). The rates available (depending on the network) include 384 kbps (6 × 64 kbps, called H0), 1536 kbps (24 × 64kbps, called H11, delivered over a T1 circuit) and 1920 kbps (30 × 64 kbps, called H12, delivered over an E1 circuit).

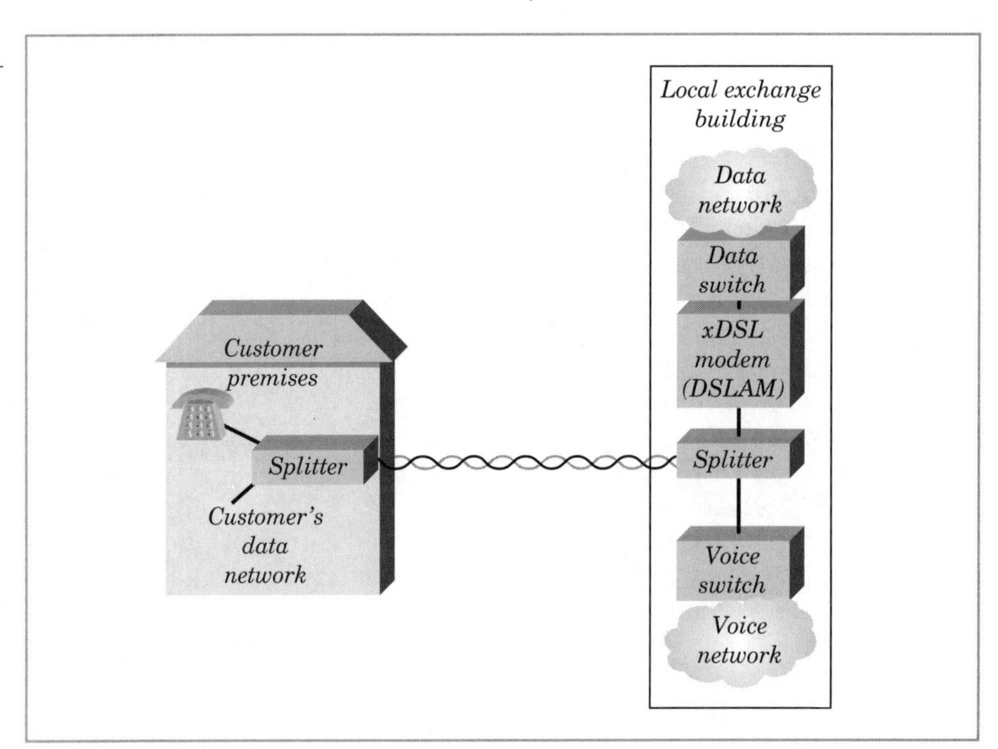

Figure 3.2 xDSL

Digital subscriber line (or loop)

Since the mid-1990s, a number of competing DSL technologies have emerged, offering various combinations of upstream data rate, downstream data rate, and range. The generic architecture of DSL systems is shown in Figure 3.2.

HDSL (*high-speed* or *high bit-rate digital subscriber loop*) uses baseband line signalling similar to ISDN BRI, but at higher data rates. By combining the capacity of two or three pairs, HDSL can carry a 2 Mbps E-1 (or a 1.5 Mbps T-1) over 3 km or so, and can thus deliver E-1 capacity. Where HDSL is used on a single pair, it is sometimes called *single line DSL* (SDSL). HDSL systems using one, two or three pairs are defined in ETSI TS 101 135 V1.4.1.

ADSL ('asymmetric', or 'asynchronous' or even 'advanced' digital subscriber loop) uses a large number of modulated carriers in the frequency range 25 kHz to 1.1 MHz, to carry 2–8 Mbps downstream (from network to user) and 64 kbps upstream, over the same range as HDSL. This is enough capacity for delivering video services (compressed), as well as Internet access. The frequency range used by ADSL leaves the 400–4000 Hz voice band free, so that the same copper pair can carry both the ADSL data and ordinary analogue voice traffic. However, a splitter unit is required at the customer premises to introduce filters to protect the data carriers from noise introduced by the voice equipment. Because of the cost of supplying, and then installing, these splitters, ADSL is being superseded by *splitterless ADSL*, alias *DSL Lite*, alias *Universal ADSL*, alias *G.Lite* (ITU-T recommendation G.992.2), which uses a more robust data signalling system.

Very high rate *ASDL* (VDSL or VADSL) offers as much as 51.84 Mbps downstream and 19.44 Mbps upstream. This is fast enough for HDTV and other very demanding applications. However, VDSL's range at these rates is only 1 km or so, and it is therefore not generally applicable as a complete local loop solution. Instead, it is used in *hybrid fibre co-ax* (HFC) solutions, where an optical fibre link reaches from the local exchange to a point within 1 km of the user.

A more widely applicable emerging technology is HDSL2, developed by the HDSL2 Consortium. HDSL2 will carry 1.5 Mbps, bidirectionally, for more than 3 km over a single copper pair (depending on wire gauge). HDSL2 is therefore very attractive as a delivery vehicle for T-1 circuits without the need for installation of new media.

There are lots of other DSL technologies, many of which are proprietary. For example: CDSL (consumer DSL); IDSL (ISDN DSL); RADSL (rate-adaptive DSL) and UDSL (unidirectional DSL).

DSL modems, I stress, work only over the short distances of the access network, and cannot work end to end across the network. Therefore, they have to connect to DSL modems or DSLAMs (*digital subscriber line access multiplexer*) in the local exchange building. From there, the traffic has to be carried by some other transmission technology.

For network users, shifting to DSL entails replacing their V.90 modems with DSL modems, as well as paying the network operator's enhanced charges. There are some efforts to offer dual-mode modems. However, as V.90 modems are now very cheap, it hardly seems worth the trouble. The shift to DSL also depends critically on the quality of the user's copper pair. In many cases (perhaps as many as 50 per cent in some networks), pairs which are fine for voice traffic are inadequate for DSL. In particular, recent aluminium pair installations do not support DSL as well as copper pairs do.

Pair-gain systems

Another approach to squeezing more capacity out of the copper access network is to use digital transmission technology to deliver additional analogue connections. Thus, as illustrated in Figure 3.3, the *digital access carrier system* (DACS), one of several pair-gain systems, delivers two PSTN connections over a single pair.

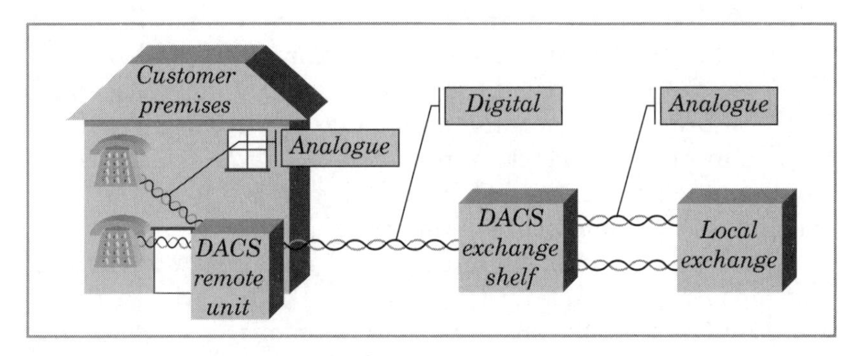

Figure 3.3 DACS – a pair-gain system

Coaxial cable

Copper also appears in the form of coaxial cables, or 'co-ax'. Co-ax is capable of carrying as much as 100 Mbps over many kilometres, using FDM and radio frequency carriers. The ITU-T recommendations for FDM on co-ax are G.332 – G.346. Although repeaters are necessary on long lines, they can be powered by a DC supply applied to the same conductors as the signal. Indeed, in the extreme case of long-distance submarine cables, several kilovolts are applied at each end, to ensure a power supply to the repeaters in mid ocean. Co-ax used to be the primary medium for telco trunks and submarine cables. Now it has been largely superseded, as a trunk medium, by optical fibre, although it is still installed, mostly for CATV and for delivery of T-1/E-1 circuits across the access network.

Optical fibre

Optical fibre is by far the most common trunk transmission medium. There are, for example, already more than 1 500 000 km of fibre in the UK, and more is being laid.

An optical fibre consists of a very thin core of glass, surrounded by a layer of glass of a higher refractive index. The two together are very fine, about 125 μm across, and so have to be protected by a plastic sheath. Several fibres are then laid together into a cable, which can be buried in ducts or run overhead. Optical fibre is so successful because it offers remarkable advantages over metallic cables. It is immune to electromagnetic interference, and so can be laid in electrically hostile environments without trouble. It does not corrode. It is not affected by water. It is cheap to make, because the raw materials for glass are some of the most abundant on the planet. It is very resistant to eavesdropping.

> Optical fibre is very resistant to eavesdropping.

Lastly, unlike copper, it is not liable to theft, except by the intellectually challenged. Copper is valuable and attractive to thieves all over the world, whether for commercial recycling or for the construction of personal ornaments. Optical fibre, on the other hand, is not valuable enough or versatile enough to have the same appeal to the nefarious. When exposed runs of fibre are laid, it is not uncommon to leave odd cut lengths lying around, so that prospecting criminals can be saved the trouble of pinching something valueless.

Digital use of fibre

Most fibre optic transmission systems are digital. They use baseband signalling, with pulses of coherent light generated by laser diodes at around 1300 nm wavelength, and detected by phototransistors. Early fibre systems used multimode step index fibre, illustrated in Figure 3.4. The central glass core was relatively wide (50–60 μm) and the discontinuity between the core and the outer glass layer was sharp. Photons entering the fibre would follow paths which would not necessarily be perfectly aligned with the fibre, but when they reached the step in refractive index, a physical phenomenon called internal reflection would cause them to bounce back into the core. To an extent, this was a success. It meant that detectable levels of light could be conveyed for reasonable distances. However, there were two serious limitations.

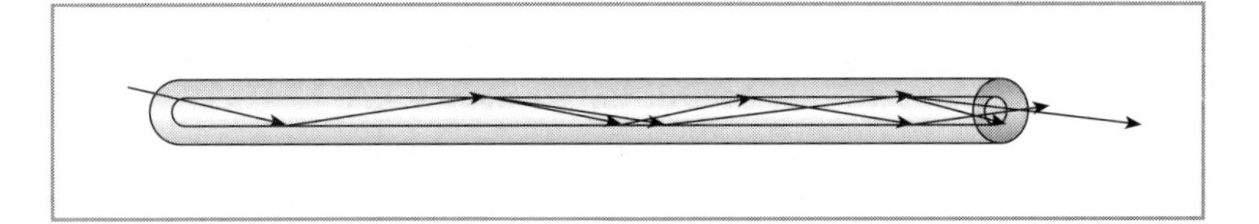

Figure 3.4 Multimode step index fibre

Firstly, the breadth of the core meant that photons could end up striking the step at quite steep angles. Internal reflection works properly only at very shallow angles, so a lot of light was lost into the outer layer. Secondly, the breadth of the core meant that although a pulse of photons would be launched into the fibre all at the same time, they would each follow a slightly different path length through the fibre, and so arrive at the end over a period of time. This pulse spreading limited the data rates achievable.

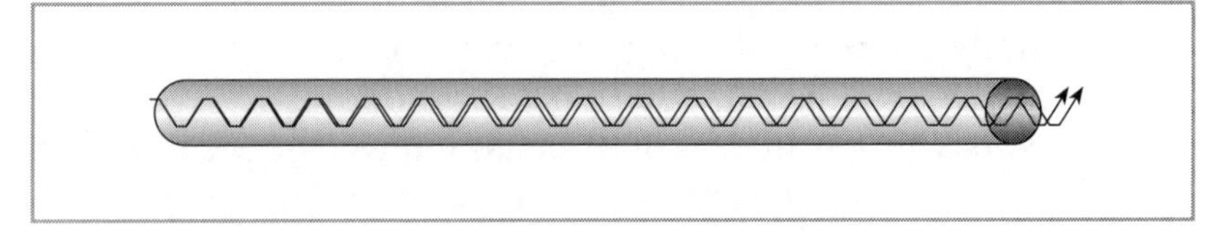

Figure 3.5 Multimode graded index fibre

This technology was superseded by multimode graded index fibre, as illustrated in Figure 3.5. Graded index means that instead of a step change in refractive index, there is a gradual change, which bends the light rays rather than reflecting them, and so avoids reflection losses. Multimode graded index can deliver 140 Mbps over 8 km. It in turn has been superseded by monomode fibre, as shown in Figure 3.6. Monomode fibre returns to a step index, but has a very narrow (8–10 μm) central core. This means that the angles of reflection are shallow and losses are thus minimized. Also the shallow angles mean that there is much less difference in length between the possible paths through the fibre, so pulse spreading is reduced.

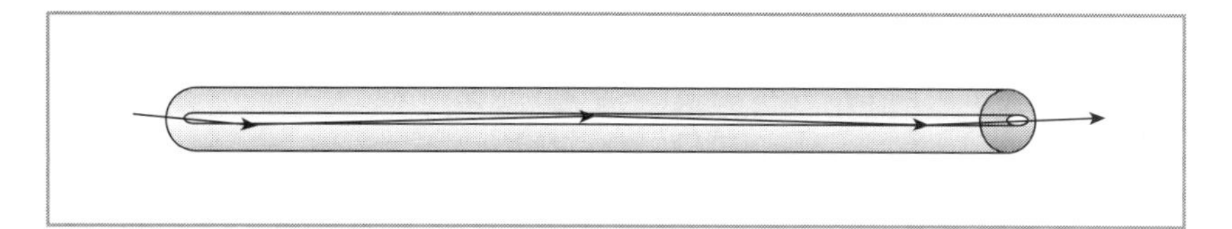

Figure 3.6 Monomode fibre

Standards for monomode and multimode fibres are defined in ITU-T recommendations G.650–G.654. Monomode fibre offers a capacity of over 10 Gbps, with 30–40km between repeaters. This is radically more than what could be achieved over co-ax, and has contributed to a drastic reduction in transmission costs. Capacity, so the mantra goes, is almost free. However, amazingly, there isn't enough of it. New applications are demanding undreamed-of transmission capacity. Many transmission networks are already suffering from *fibre exhaustion.*

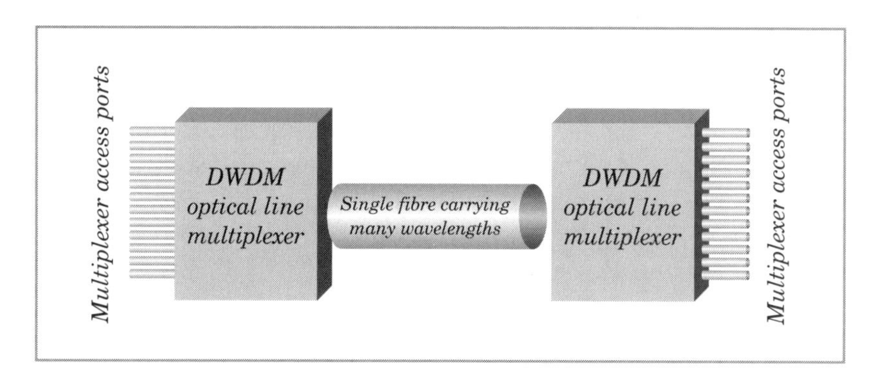

Figure 3.7 DWDM

One solution would be to lay more fibre, but that is very costly in civil engineering work. Another approach would be to increase the signalling speed. However, pulse spreading and transistor switching rates make it difficult to go much beyond 10 Gbps. The practical solution is to use more than one colour of light (we should say wavelength really, because at 1300 nm, the light used is just outside the visible spectrum).

The technology for simultaneously signalling with several wavelengths down a single fibre is called *wavelength division multiplexing* (WDM) (*see* Figure 3.7). Recent improvements in filter[2] and laser technology[3] have allowed higher numbers of distinct wavelengths to be used, under the banner of *dense WDM* (DWDM). DWDM line systems offer to provide an aggregate capacity of 128 × 10 Gbps, that is to say, over 1 terabit per second, overall, down a single fibre.

The many channels that a DWDM transmission system provides can be used independently, for SDH, ATM, IP or whatever. While this is in some ways a good thing, it is also a new challenge for network management systems, which for years have been developed from the idea that each higher-level transmission system will have its own physical medium. The fact that a fault in one fibre could affect several quite different transmission systems can be quite difficult to manage.

Analogue use of fibre

Not all fibre optic transmission systems are digital. In a small minority of networks, especially in CATV distribution networks, optical fibres are used to carry analogue radio-frequency signals. This is, however, difficult and costly. To faithfully reproduce the analogue waveform requires laser sources with a linear response. These are physically big and expensive, and they need frequent adjustment. Analogue transmission over fibre is likely to dwindle away as all-digital CATV networks are deployed.

Many network operators have made efforts to extend their use of optical fibre into the access network.

[2] The filters used are called Bragg gratings.

[3] In particular, narrowband lasers, which emit only a closely restricted range of wavelengths.

Optical fibre in the access network

Many network operators have made efforts to extend their use of optical fibre into the access network. While this brings benefits in higher capacity and higher reliability, it has the disadvantage of immense civil engineering costs if the copper network is replaced all the way to the customer's premises. A range of compromises have evolved, including FTTH (*fibre to the home*), FTTB (*fibre to the building*), FTTK/FTTC (*fibre to the kerb/curb*) and FTTCab (*fibre to the cabinet*). These are illustrated in Figure 3.8.

Passive optical networks

All of these technologies can use (and increasingly do use) a new family of access network transmission technologies called *passive optical networks* (PONs), alias *telecommunication over PON* (TPON) or *optical access networking* (OAN).

The key idea of PON is to avoid having active components in the access network, not only because of their capital cost but because they would have to be looked after and also probably replaced as obsolete before long. A PON, as shown in Figure

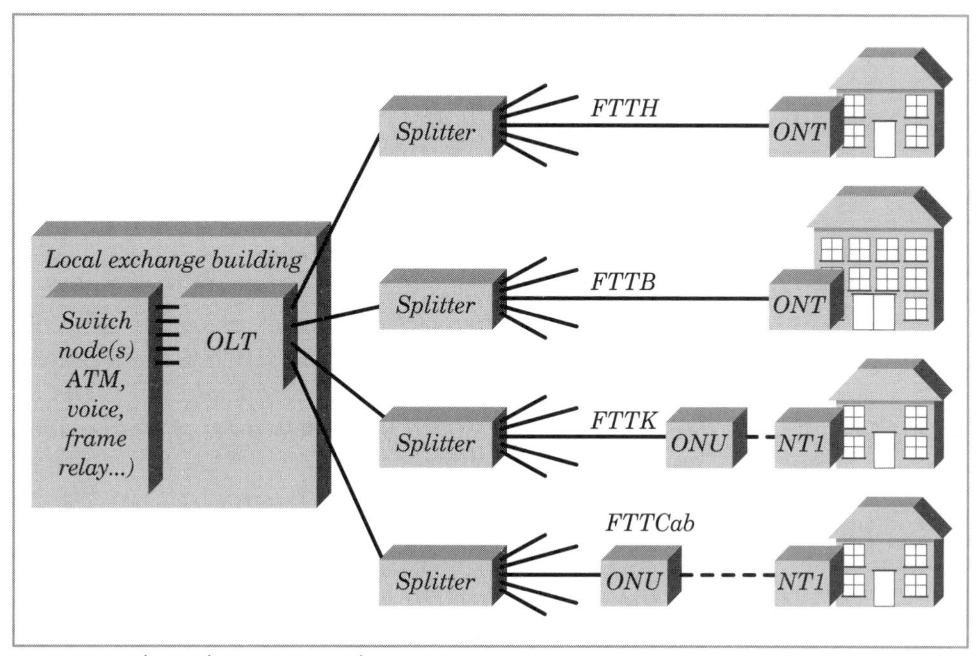

Figure 3.8 Fibre in the access network

3.8, has a fibre optic star network extending from a distribution node (either an SDH node in an existing fibre access network, or local exchange building itself) out to the customers' premises. The nodes of the tree use cheap passive optical splitters. So the optical signal from the exchange end goes to all the users' end-points, and the signals from the customers are combined as they move towards the exchange.

In the case of FTTCab and FTTK, the optical network terminates at *optical network units* (ONUs), which convert the transmission to copper pair for delivery to an NT in the customers' premises. With FTTCab, this happens in a neighbourhood cabinet up to 1 km from the customer. In FTTK, it is done in a smaller street cabinet within 200 metres or so of the customer.

FTTH and FTTB both carry the optical network into the customer's premises. In these PON variants, the termination point for the fibre network is called an *optical network termination* (ONT). While this can increase installation costs, it means that there is no electronic equipment in the access network to maintain. Also, upgrades to the network capacity can be made by simply changing the OLT (*optical line terminator*) and the ONT, which protects the network operator's investment in the access network.

Multiplexing and access management over PONs can be achieved in a number of ways. Firstly, WDM can be used, so that each wavelength is allocated to one user or a group of users. Often a user will not require a whole wavelength's worth of capacity, so the wavelengths will each be split between users, typically by using an approach supporting ATM.

As Figure 3.9 shows, in the downstream direction, TDM is used to interleave ATM cells destined for all the users on the fibre star. The ONTs at the users' premises pass through only the cells that are addressed to that user. In the upstream direction, the ONTs have to be synchronized so that their cells do not collide. A synchronization process called ranging, between the OLT and the ONTs, achieves this.

ATM PON standards promoted by the *full service access network* (FSAN) group support operation at 155 Mbps upstream, and as much as 622 Mbps downstream (shared out between all the users). The presentation of this capacity at the users' premises may be raw ATM, ISDN BRI, T-1/E-1, frame relay or other formats, depending on the capability of the ONUs.

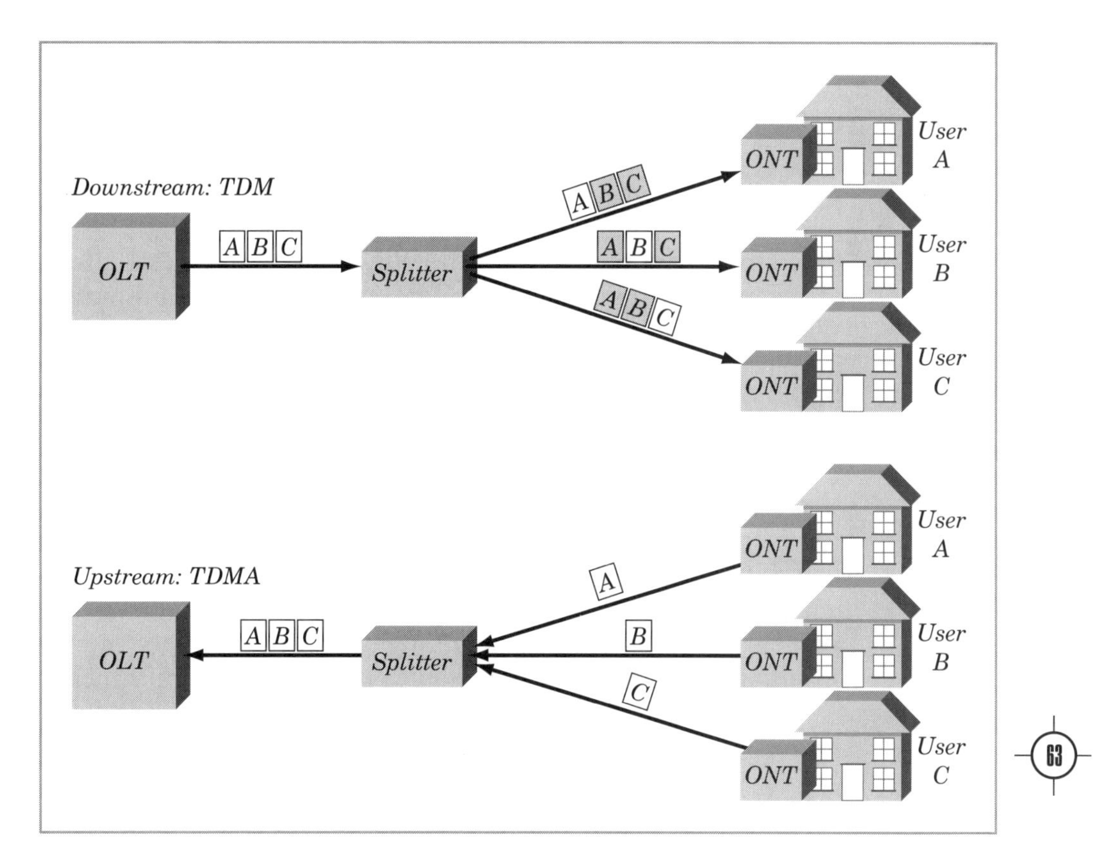

Figure 3.9 ATM PON media access control

Blown fibre

Outside the customer premises, the mechanics of fibre networks are not grossly different from those of copper networks. Once inside a business property, however, blown fibre can be used. The building is prepared by installing a system of small-bore ducts. Then, when a new fibre needs to be installed, it is fed in at one end of a duct, together with a forced air flow. The draught carries the fibre through the duct to its destination. This droll process is justified because it makes fibre installation a one-person job, thus saving on field staff costs.

Dark fibre

Not all the owners of fibre networks bother to connect transmission or switching equipment to them. Some companies dig trenches, install ducts and fibres (between likely locations), and then sell the use of it to network operators. Such raw fibre is called 'dark fibre' or occasionally 'grey fibre'.

All-optical network technology

As we noted above, there is a practical limit to how fast a fibre can be made to go if it has to interface to electronic equipment. While electronics continues to get faster, there is likely to be a hard limit at some time, because of simple issues such as capacitance of circuit elements. There is therefore much interest in developing all-optical network systems. The following are three all-optical devices which exist now and which illustrate the possibilities of all-optical technology.

- *Erbium doped fibre amplifiers* (EDFAs) are all-optical amplifiers, which are electrically powered but which do all the signal amplification using light. These introduce the possibility of long fibre runs with optical repeaters, which would not be subject to the same data rate limitations as electrical repeaters.

- Optical switches are already well enough advanced to support *optical line section protection*, as shown in Figure 3.10. Here an optical switch (which does not need to understand the transmission formats used, or even the line rate) changes automatically to a fallback route if the primary route fails.
- Passive *wavelength space switches* are another interesting kind of optical circuit switch. They are solid objects constructed of transparent material, with shape and

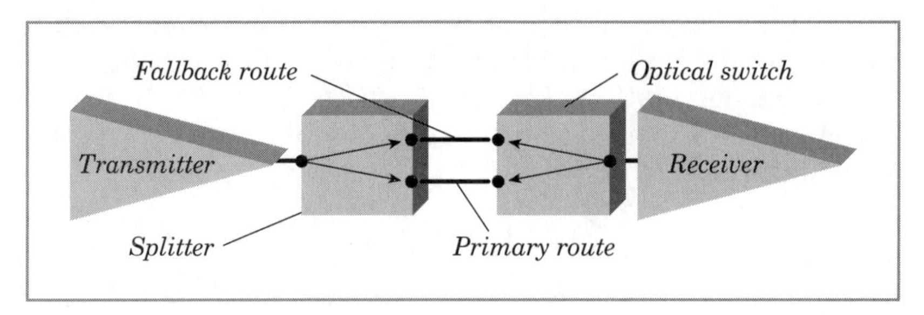

Figure 3.10 Optical line protection

refractive index designed so that the light from any of the inputs will go to any one of the outputs, depending on its wavelength. Thus circuit switching can be achieved by modifying the input wavelengths at their source, through use of tuneable lasers. Devices with 128 inputs and 128 outputs have already been built.

Lastly, before we leave optical fibre transmission, it is worth noting (for calculating transit delays) that light in optical fibres does not actually run at the standard 3×10^8 ms^{-1} speed of light in a vacuum, but somewhere around a slug-like 2×10^8 ms^{-1}.

Free-air optical transmission

Not all optical transmission systems rely on fibre to carry the light. Infrared LAN technology is discussed in Chapter 4. Of more significance for telecommunications networks is the recent improvement of free-air optical transmission systems. The problem with this technology has always been the weather. The first weather problem is interference from precipitation, or birds, for that matter. By using lenses to spread the light beam, the interfering effect can be minimized. The other problem is that the wind can sway the buildings to which the terminals are attached, making tiny changes to the signal path length, and thus introducing jitter. Now there are tracking and servo systems good enough to compensate for this. The results of these improvements are systems that can carry 2.5 Gbps and more, over a small number of kilometres. However, they are still weather dependent, as fog can scatter the beam and make them inoperative.

The problem with this technology has always been the weather.

CATV

CATV is a peculiar mix of broadcast technology, transmission technology, co-ax, fibre and twisted pair. While none of the technologies that it uses is unique to CATV, there is so much CATV around that it is worth getting to know how it all hangs together. When Alice asked Humpty Dumpty what 'Glory' meant,[4] he replied that it meant just whatever he paid it to mean. CATV is like that. It could stand for *community antenna television*. Or it could be *community area television*. Or it could be *cable television*. Take your pick.

Early CATV systems happened because people (largely in the US) wanted better TV reception than they could get from their household antennas. A large antenna would be erected in the centre of the area (at a building which came to be called the 'head end'), and the signal from it would be amplified and distributed over a unidirectional (outwards only) coaxial cable network, to people's houses, as Figure 3.11 illustrates. From this, it was a small step to equip the head end with a satellite TV receiver, and a few industrial-grade video players, and use the considerable spare RF bandwidth of the co-ax network to offer CATV subscribers a wide range of programmes.

From these simple beginnings, CATV networks have evolved in a number of directions, as the following figures show. It has not been a uniform evolution, and any CATV network shows the influence not only of the current market trends and technology but also of its own history.

When optical fibre became available, it became cheaper to lay fibre for the first few kilometres from the head end, and then convert to co-ax for delivery into homes. Such networks were

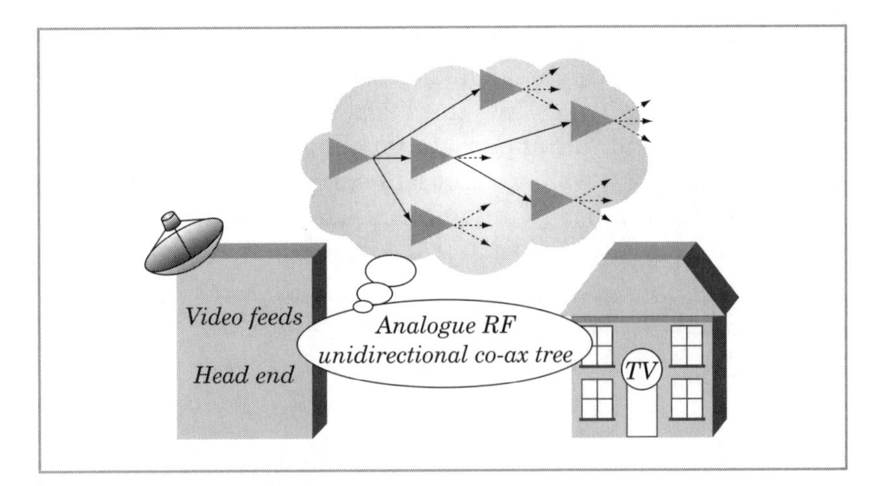

Figure 3.11 Early CATV network

[4] Carrol (1872).

(and are) called *hybrid fibre co-ax (*HFC). Early HFC networks used RF analogue signalling over the fibre, as discussed above and shown in Figure 3.12.

To better serve or exploit their customers, most CATV

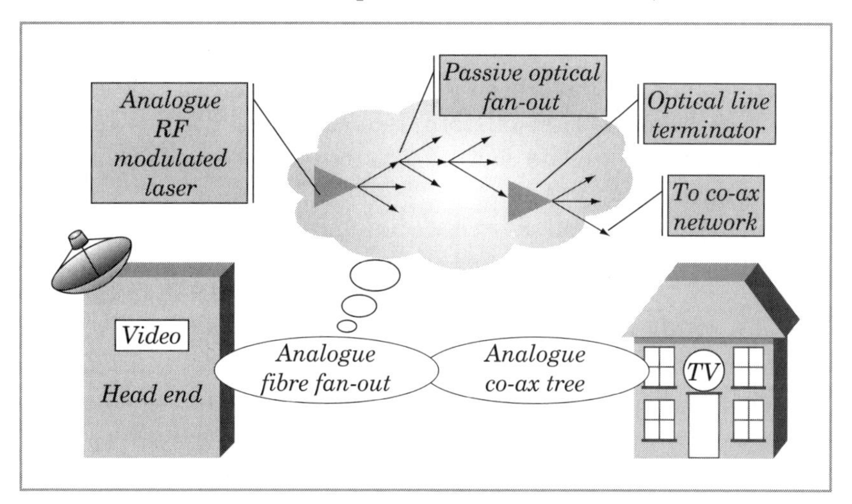

Figure 3.12 HFC network

network operators segregated their channels into basic ones that all their customers could receive, and premium ones that customers could receive only for extra payment. A refinement of this was to exact payment for specific events (*pay per view* – PPV). Control over channel access was achieved by scrambling the premium channels in a *conditional access* (CA) system at the head end, and equipping each home with an *integrated receiver decoder* (IRD), otherwise known as a set top box. This arrangement is shown in Figure 3.13. The CA system could

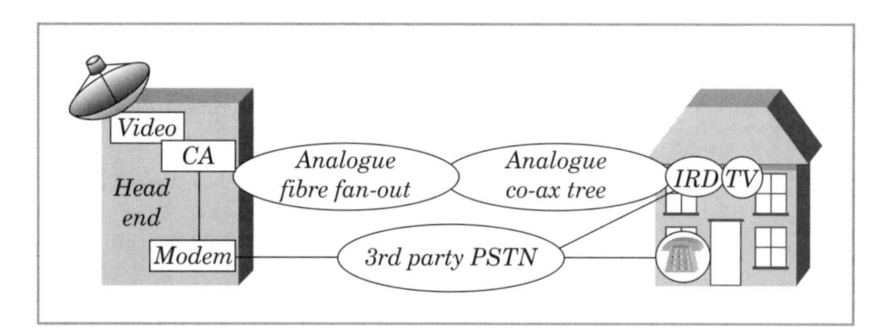

Figure 3.13 Conditional access

control which IRDs could decode which channels by distributing the necessary decoding keys, over spare RF bandwidth, with associated IRD addresses (each IRD has a unique address).

Customers could request premium services by calling the head end over the (entirely separate) PSTN. Or for simpler control, a modem integrated into the IRD could call a modem at the CA system, so that customers could order PPV events by pressing buttons on the remote controller for their IRD. This is called *impulse pay per view* (IPPV).

Recently, spare capacity in analogue CATV networks has been used for providing customers with Internet access. The *data over cable service interface specification* (DOCSIS – produced by a gang of cable operators) covers several aspects of delivering data services to customers, including a specific interoperable *radio frequency interface* (RFI) for *cable modems*. A cable modem at the customer's end (tee-ed into the incoming co-ax) connects to a *cable modem termination system* (CMTS) in the head end. The DOCSIS RFI uses spare bandwidth in the RF network to deliver downstream data rates of 20–40 Mbps and upstream rates of 160 kbps–10 Mbps. Downstream data is split between customers by TDM and FDM. Upstream access is shared through TDMA or a form of contention management comparable to Ethernet.

Implementing cable modem systems has been attractive to the CATV operators because the early availability of cable modems has enabled them to offer high-capacity Internet access before the copper-pair telcos could deploy DSL systems. However, network upgrades are often required. Many CATV networks have RF amplifiers and laser systems facing downstream, and no amplification or signalling on the upstream signal path. In such cases, substantial investment and network redesign has been necessary.

Another possible use of spare RF bandwidth is to carry the voice telephony signals, via a box in the home called a telco modem, which modulates analogue voice signals onto otherwise unused RF carriers. Where CATV networks were being installed in countries where competition for telephony services was allowed, CATV operators installed telephony networks in parallel with the co-ax RF network, as shown in Figure 3.14. The CATV operators saw this as a chance to move up-market (in the early days of CATV, its market focus was on low-income groups), as well as a chance to win TV customers through the appeal of their telephony offering.

The head end would contain an ordinary local telephony switch, connected either to the CATV company's own trunk network (where licensing allowed) or to a long-distance carrier. The switch would be served by a brand new local loop using SDH, Sonet or PDH out to each neighbourhood (say of 500 homes), where a demultiplexer and codecs converted to copper-pair analogue transmission for delivery into customers' homes. The copper pair would be wound around the RF co-ax where they both left the slot box in the pavement outside a customer's home. This gave the cable entering the home a spiral appearance, whence it got the name of 'umbilical'; on a similar theme, it is also called 'Siamese'. It is essential to note that at this stage, the RF TV and telephony networks are technically distinct, related only by having a common physical infrastructure, ownership and customer base.

This gave the cable entering the home a spiral appearance, whence it got the name of 'umbilical'.

The opportunity for integrating the TV and telephony

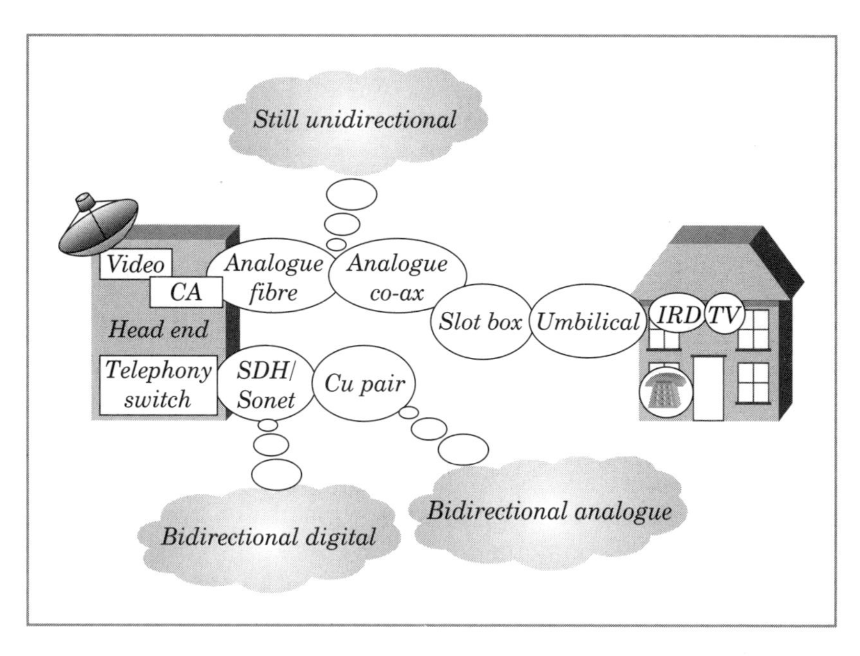

Figure 3.14 Parallel telephony network

infrastructure has arrived recently, with digital TV, as shown in Figure 3.15. Digital TV signals can be transmitted over the SDH/Sonet network, so the old analogue RF fibre network can be scrapped. Because the quality of digital signals can be maintained over any distance, local head ends can be reduced to network nodes, and all the programme content generation and access control can be centralized. By converting to co-ax close to the home, the co-ax can be used without repeater amplifiers, and so can be used bidirectionally to carry the digitally modulated TV, telephony and other data. The umbilical and copper-pair network can also be dispensed with, and the telephones can be connected via the IRD. In real life, few networks have got this far, but some have, and this is where they all will be heading.

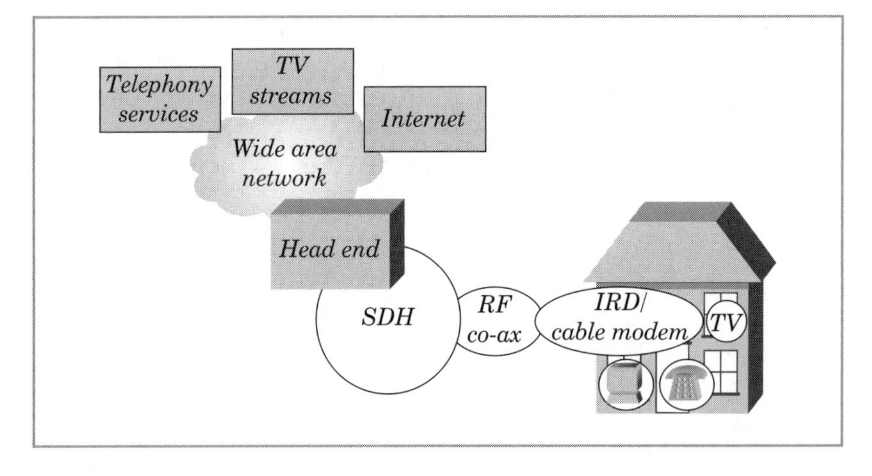

Figure 3.15 Integrated digital TV and telephony network

Fixed-terminal satellite

The first telecommunications satellite, Telstar, was launched in 1962. There are now a number of satellite systems offering international telecommunications capacity. With just three satellites in geostationary orbit (36 000 km high) above the equator, it is possible for a satellite network operator to cover the whole planet except for the north and south poles.

An example of a telecommunications satellite system is the one operated by Intelsat. It operates uplinks on 6 GHz and downlinks on 4 GHz. The system uses 32 metre diameter earth station antennas, which connect into national carriers' networks.

A more accessible satellite technology is *very small aperture* (VSAT), which works with much smaller, almost portable antennas (typically 2–3 metre diameter in the US, where a 4–6 GHz radio band is used, and 1.8–2.4 metres in Europe, using 12–14 GHz). VSAT is used for quick provision of high-capacity (multi-megabit) services to businesses in remote locations, where it would take a long time to install a new fibre link. Television field crews also sometimes use VSAT to transmit news back to their studios. For voice telephony, a serious drawback to satellite transmission is the 250 ms round-trip delay that is incurred by the high geostationary orbit. This is long enough to make conversation awkward. Now, most international traffic is carried by undersea optical fibre systems, which are much shorter and do not present that problem.

Microwave radio

Microwave radio operating above 1 GHz can be used for point-to-point links providing a capacity of the order of 100 kbps over distances as great as 50 km. Microwave installations are often very visible structures, featuring tall towers on high ground, ornamented with circular dishes and trapezoidal horns. Many countries established very large microwave networks before optical fibre made them uncompetitive in many situations.

The advantages of microwave transmission systems are:

- low capital cost, with no continuous media to provide;
- easy installation, without having to access the whole route;
- capacity for crossing difficult terrain such as mountains, rivers or bogs, which may be inaccessible to cable-laying equipment;
- not liable to disruption by events at ground level along the route (for example, road digging).

The disadvantages of microwave include:

- the need to plan the network as a series of straight line hops between repeater stations which are in direct line of sight of each other;
- the need to supply power to repeater stations which are often in difficult locations, and the need to visit them for maintenance;
- a major visual impact on the landscape;
- some deterioration in transmission quality in adverse weather.

Another problem in microwave communications, which in fact applies to any wireless technology, is multipath attenuation (also called Rayleigh fading). Figure 3.16 shows a wireless link between two stations. Three of the many possible signal paths between them are shown:

- the direct path;
- a path which bounces off the terrain;
- a path which bounces off discontinuities in the atmosphere.

The several signals will all have different path lengths, and so, when they arrive at the receiving station, may arrive out of phase with each other. If this happens, they will tend to cancel

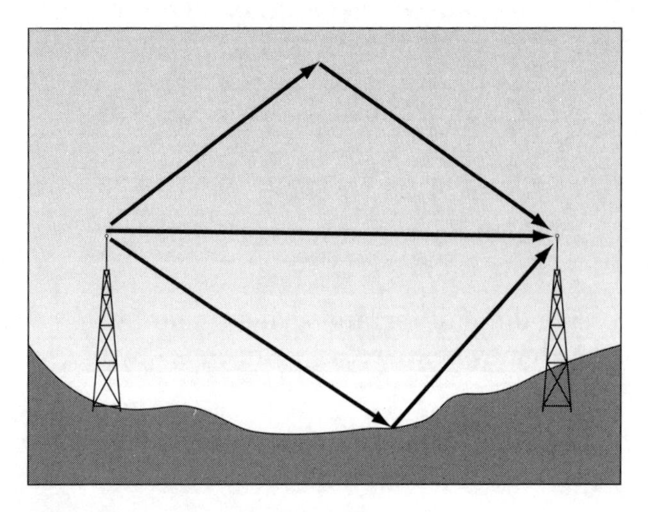

Figure 3.16 Multipath attenuation

each other out, and the resulting signal will be seriously weakened. If an FM signalling system is employed, some frequencies will be affected worse than others, and so the received amplitude will fluctuate. Multipath attenuation is difficult to model, and so may not become evident until the radio stations have been built.

While microwave used to be many networks' major trunk medium, it is now very largely superseded by optical fibre. Microwave persists, however, where cable laying is impractical; for example, in island areas, and when new operators wish to set up a network faster than the civil engineering processes of fibre laying will allow.

While microwave used to be many networks' major trunk medium, it is now very largely superseded by optical fibre.

Fixed-terminal wireless local loop

Sometimes laying copper or optical fibre to a subscriber is impractical because of installation costs, installation delays, planning regulations, distance, or risk of theft. In such cases, *wireless local loop* (WLL), also called *radio in the loop* (RITL), *radio local loop* (RLL) or *fixed radio access* (FRA), can be used. WLL, at its simplest, can be done by using mobile-terminal

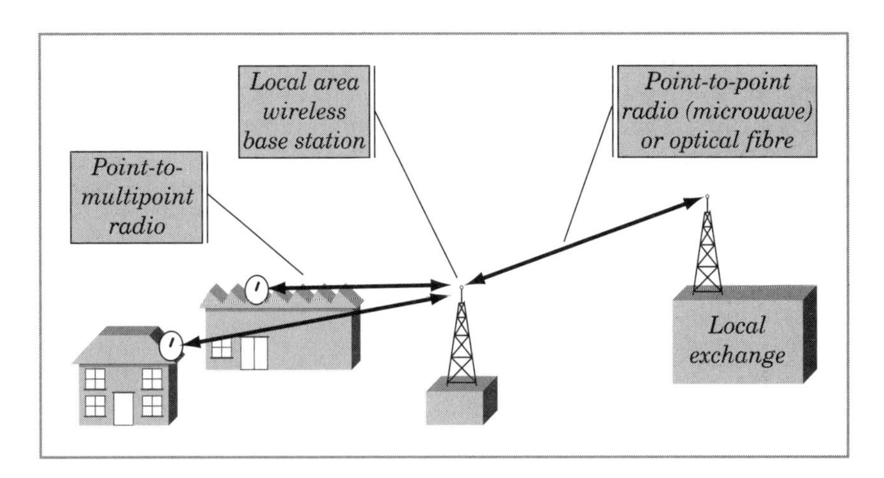

Figure 3.17 Wireless local loop

transmission technologies such as GSM, D-AMPS or DECT, without all the associated mobility management equipment. For example, GSM is used in some countries to operate fixed call boxes. However, there are also a number of specifically developed WLL technologies. Many of these are proprietary, but standardization efforts appear to be concentrating around one of them, LMDS.

The main components of a WLL system are shown in Figure 3.17.

LMDS

LMDS, the *local multipoint distribution service*, is a wireless technology operating in the 28–38 GHz frequency band. LMDS is effective over a radius of 2–8 km, depending on the terrain, the climate (it is particularly susceptible to fading in heavy rain), and the required transmission quality. LDMS provides up to 155 Mbps downstream and 55 Mbps upstream, per subscriber, with a total downstream capacity of around 1 Tbps per transmitter.

Signalling between the local exchange and the customers' units uses ATM over QPSK (quadrature phase shift keying) or QAM, and the customer can be offered an assortment of ATM, Ethernet or T-1 network interfaces.

Within LMDS, multiple access is achieved by a combination of FDMA and TDMA. The available spectrum is divided into 20 MHz channels, and these can be re-used in a cellular architecture. Channels can also be re-used in orthogonal polarizations – that is, a channel can be used in one area in vertical polarization, and in an adjacent area in horizontal polarization, without interference. Within each channel, the downstream signalling uses TDM to collate the signals for several users. Upstream, TDMA or sometimes FDMA is used.

LMDS standardization is being pursued by a number of organizations, including the ATM Forum, DAVIC, ETSI and the ITU.

The high frequencies used in WLL, and LDMS in particular, make it particularly dependent on a clear line of sight between the base station and the users' antennas. In built-up or hilly areas, this can be hard to predict, and so it can be difficult for a network operator to be certain whether they can provide service to a prospective customer.

High-altitude platforms

High altitude platforms (HAPs) are an emerging technology at the time of writing, and their commercial viability is not yet established. The idea of a HAP is to use a relay of aeroplanes, or a balloon, to position the base station for a point-to-multipoint wireless system, with associated switching equipment, in the air above a business area or city.

Electrical mains

Almost every home in the developed world has a 110-volt or 220-volt AC electrical mains connection to the national electricity distribution network. Homes without mains electricity are fewer than homes without telephone lines. Superficially, the electricity distribution network appears as if it might offer a way of delivering communications services, with a very low entry cost, and bypassing the owner of the regular access network.

There have been a number of technological initiatives by equipment manufacturers in this area. Their success has been mitigated by a number of common difficulties:

- the mains distribution network, from the local substations to the 50–200 homes that each one serves, is electrically a bus, with all the homes connected to the same pair of conductors;
- this means that the capacity of that local network (typically 1 Mbps) has to be divided between many users, by some sort of access control;
- the transformers in the substations prevent direct signalling from the distribution backbone into the local networks; in practice, a conventional optical fibre network is used to reach the substations;
- the transmission environment is electrically noisy, and bit error rates are therefore unacceptably high for circuit mode telephony, and perhaps even for VoIP (voice over IP);
- there has to be a ‘conditioner’ installed in each home, to separate the signal from the mains power.

At present, it appears that DSL technologies will deliver much higher capacity, using technology that is much better developed. Unless the issue of access to the regular access network becomes paramount, mains transmission appears unlikely to take off.

Further reading

The basics of this subject are covered well in Flood and Cochrane (1995) For more detail on developments in access network technologies, see *BT Technology Journal* (1998) and Fenton and Sipes (1996).

4 Mobile-terminal wireless transmission

The 'cellular' trick

Wireless transmission technologies that are designed to cope with the user terminals moving around can be split into two kinds: cellular and (for want of a better word) non-cellular. The cellular idea, which is both simple and clever, goes something like this.

Suppose the authorities have allocated you 2 MHz of bandwidth to use in your territory by the authorities. The transmission technology that you have chosen uses FDMA, with, say, 20 kHz per channel. So, if you erect one huge transceiver in the middle of your territory, using all of the available bandwidth, you can have 100 wireless channels, and thus 100 users simultaneously making calls on your network.

You might be a bit smarter, and replace the one big transceiver with lots of little ones, spread out over the territory. Then the mobile terminals can use lower-powered transmitters so that your customers won't fry their brains. You could call the area around each one a 'cell', and you could try to re-use your 100 channels in each cell. But this still doesn't work well, because of the way that radio waves propagate. If the transmitters are powerful enough to cover their own cells, they will inevitably leak signal into adjacent cells, which will interfere with the signals there, and make the channel-re-use scheme a failure.

If the transmitters are powerful enough to cover their own cells, they will inevitably leak signal into adjacent cells.

But suppose instead that you divide up the 100 channels into five groups of 20 channels. You build lots of transceivers all over your territory. To each transceiver's cell, you allocate one of those groups, and you arrange things so that no two adjacent transceivers use the same group. Then signal leakage between adjacent cells can only be of frequencies that are not being used in the affected cell. In each cell you can have 20 simultaneous calls. If you have, say, 1000 cells, you can have 20 000 simultaneous users. The total number of users you can have simply depends on how many cells you are prepared to divide your territory into.

That is the cellular trick. It has been essential for mass-market FDMA wireless services, because bandwidth is an inescapably limited resource, which is in short supply already. Cellular frequency re-use designs are also sometimes used in fixed-node wireless networks (for example in LMDS), but the technique has come to the fore in mobile communications networks since the late 1970s. Before that time, it was not practical to equip either the network or the mobile terminal with enough computer power to organize the necessary frequency changes as the user moves from cell to cell.

Non-cellular systems

Paging

Paging systems offer a unidirectional (push-only) service, which can include the following variants:

- alert-only;
- alert, with immediate delivery of a voice message;
- alert, supported by a voicemail system which the pager user can call back (on a separate voice network) to retrieve a message;
- alert, with a short text message.

Recently 'pagers' have been introduced offering bi-directional facilities. However, these are based on the cellular network technologies presented below, which are quite different from ordinary pager technology.

It would be satisfying to offer the reader a neat, compact and comprehensive account of the radio interfaces used in pager networks. However, that is not possible. There are many systems (which are generally incompatible). Some use *long-wave* radio; others use *ultra-high frequency* radio. Signalling over the radio carrier can be by FSK or by voice-band MF tone signalling. Most paging systems are terrestrial, but some are satellite based. Of the many, many systems, probably the most widespread is *POCSAG* (gloriously named after the Post Office Code Standardization Advisory Group), which offers 2.4 kbps signalling.

With so many paging network technologies, it is hardly surprising that roaming between networks (i.e. when you buy a pager to work with network A, and then move into network B's territory, and hope that your pager will keep on working through some tie-up between the networks) is piecemeal, ad hoc and inconsistent. One example of a current roaming protocol (by which the two networks pass the paging messages between themselves) is TNPP (the *telelocator network paging protocol*). The problem is, there are loads of others.

Recently there have been attempts to get roaming arrangements into a more orderly state. ERMES (*European radio messaging system*) offers ETSI standards for roaming agreements within Europe, and is likely to be implemented in any new pager networks. However, given the increasing functionality and cheapness of cellular systems, it may be that paging systems have already passed their growth phase.

PMR and PAMR

PMR can stand equally well for *private*, or *professional*, *mobile radio*. A number of national-scale organizations (for example, some large roadside assistance organizations) have their own analogue radio networks, with fibre transmission systems feeding numbers of transceiver stations.

Alternatively, an organization can subscribe to the services of a PAMR (*public access mobile radio*) system, which allows its subscribers to share access to *common base stations* (CBS) and transmission infrastructure.

Both PMR and PAMR can offer two features that are not found in most kinds of communications networks: group call and direct call. Group call means that one user (for example, a police dispatcher) can simultaneously speak to a group of mobile users. Direct call means that users can call each other directly, without having to go via any fixed network infrastructure. This feature (at the cost of powerful and therefore large handsets) allows PMR and PAMR to support users in extreme situations where normal mobile networks cannot be relied on.

First-generation (analogue) cellular

The first generation of cellular mobile networks used analogue transmission. A litter of similar but incompatible technologies emerged in the mid-1980s:

- TACS (*total access communications system*) and ETACS (*extended TACS*) in the UK, alias JTAC in Japan;
- NMT-450 (*Nordic mobile telephony*) and NMT-900 in Scandinavia;
- NMT-F, RTMS and RC 2000 in France;
- C-Netz in Germany and Austria;
- NTT mobile and HICAP in Japan;
- AMPS (*advanced mobile phone service*) and NAMPS (*narrowband advanced mobile phone service*) in the US, and others.

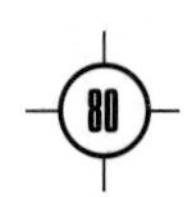

First-generation systems had some serious drawbacks. The transmission technology used bandwidth inefficiently, limiting the possible number of users per cell. Calls were easy to eavesdrop on. Worst of all, the mobile devices transmitted their identifying information without encryption. It was (and still is) easy to buy devices which will intercept this information and enable fraudsters to make 'clone' copies of the terminals, which they can then sell. The calls made on the cloned phone are charged to the victim, until somebody notices. This led to interesting developments in fraud-busting techniques by the network operators, but it remains a problem.

First-generation network services were restricted to voice,[1] with poor indoor coverage, phones as heavy as housebricks, and expensive tariffs aimed at business users only.

[1] Except for modem data calls, and the 19.2 kbps cellular digital packet data (CDPD) overlay of AMPS (essentially a modem service using some spare carrier capacity).

As mentioned above, some first-generation networks exploit the capability of their signalling channels to offer paging services, with the possibility of bi-directional working (i.e. from the pager as well as to the pager).

Second-generation (digital) cellular

The second generation of mobile cellular networks used digital transmission systems. As well as voice, they offered (initially limited) data services. Coverage could be achieved indoors through microcells. The phones were small, and services were cheap enough for the mass consumer market.

The basic GSM Phase 1 system

There are several second-generation, digital cellular network technologies. One of them, GSM, is so widespread that I use it here as an example of how that kind of thing works.

Once upon a time, GSM stood for Groupe Speciale Mobile. This matched its origins as an essentially European effort to achieve a common, interoperable standard for second-generation systems in Europe. Later, to support its use (or enhance its credibility) as a global standard, it was renamed the *global system for mobile*. The GSM standards are managed by ETSI, and have been released in three waves: Phase 1, Phase 2 and Phase 2+.

GSM Phase 1 features

Ordinary POTS-like service.

Call forwarding (all calls, or on busy, or on no answer, or if unreachable).

Call barring (either for outgoing or incoming calls).

Global roaming between GSM networks.

GSM Phase 2 features

Short message service (described below).

Multi-party calling.

Call hold.

Call waiting indication.

Circuit switched data (described below).

Fax transmission.

CLI presentation.

Advice of charge.

Cell broadcast (described below).

GSM Phase 2+ features

AMR, HSCSD and GPRS.

CAMEL.

SIM toolkit.

Routing optimization.

Call interception.

(Phase 2+ is merging into the developing world of third-generation cellular networks.)

Radio interface

The GSM radio interface (the Um reference point) was originally specified to use frequencies around 900 MHz.[2] The primary access management system is FDMA, with RF carriers spaced at 200 kHz. Slow frequency-hopping between these

[2] But, as noted later, it has been extended to a number of other frequency bands.

carriers is used to achieve a degree of resilience against fading, which can be severe at these frequencies.

Within the carrier structure provided by the FDMA, TDMA is used[3] to provide eight traffic channels per carrier, which typically carry voice traffic encoded in 13 kbps via *regular pulse excited linear prediction coding* (RPE-LPC). The traffic channels are allocated only when there is a call, or call attempt. Also, to minimize crosstalk and conserve power, *discontinuous transmission* (DTX) is used; that is to say, the transmitters operate only when there is actually some speech to carry.[4] The traffic is also encrypted using a subscriber-specific ciphering key. The TDMA structure also provides a number of control channels, which are monitored by all live mobiles, whether they have a call or not.

GSM networks typically use a seven-cell repeat pattern. In public networks, cell radii range from about 1 km to 30 km. GSM is also used in some corporate networks as an in-building system, with micro-cells (or pico-cells, the terms are not clearly distinguished) of just a few tens of metres across. Signal propagation is difficult to model, and so cell locations, sizes,

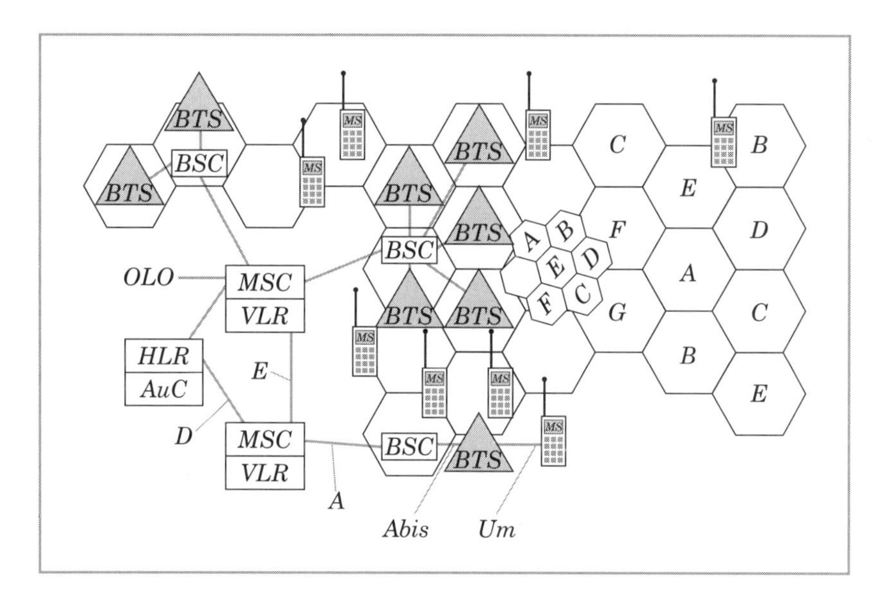

Figure 4.1 GSM network components and reference interfaces

[3] GSM is therefore sometimes classed as a TDMA system.

[4] The receivers simulate background noise during these gaps in the transmission, so the listener does not notice.

transmitter powers and carrier allocations are developed iteratively, in response to network availability measurements.

Figure 4.1 illustrates a part of a GSM network, showing the major component types.

- A *mobile station* (MS) comprises the *mobile terminal* and a *subscriber identity module* (SIM). The mobile terminal (phone) does the radio signalling (0.8 W transmitter power, or sometimes more) and audio processing. The SIM is a removable smart card with a microprocessor, and small amounts of EEPROM and RAM. The SIM contains all the information that is specific to the user, such as their IMSI (*international mobile subscriber identity*) and authentication key. It is therefore possible for a user to change from one phone to another by swapping the SIM.

The SIM contains all the information that is specific to the user.

- A *base transceiver station* (BTS) is a radio mast and its associated RF electronics. A BTS may serve a single cell, or it may be placed at a corner between cells, to serve as many as three cells, using directional (phase-array) antennas. Backhaul from a BTS to its BSC (base station controller) is done over either a fibre optic link or a point-to-point microwave link.

- A *base station controller* (BSC) concentrates the traffic from up to 100 or so cells, and does most of the control of the RF signalling. It manages allocation of traffic channels for data of speech, and the airside signalling to the mobile device. It also converts the RPE-LPC voice encoding to the usual a-Law PCM used in the rest of the network (in a *transcoder rate adaption unit* or TRAU).
- A *mobile switching centre* (MSC) is in many respects just an ordinary circuit switch, like a fixed-network local exchange. The differences are in its call control logic, which is modified to handle the mobility of the users through interaction with HLR (*home location register*) and VLR (*visitor location register*) systems. A GSM network may have many MSCs, connected in a trunk network like any other circuit switched telephony network. The MSCs that connect to other networks have a special role in handling

incoming calls to mobile users, and are called *gateway MSCs*.

The call control signalling between the MS and the MSC is close to the ISDN protocol Q.931, while the MSC's signalling to the HLR uses the SS7 mobile application part (MAP) defined in ITU-T Q.1061–1063. Q.931 and SS7 are explained in Chapter 6 of this book. The facilities provided by MAP (which runs on top of SS7 TCAP) include:

- transfer of subscriber information from HLR to VLR;
- location updating (from VLR to HLR);
- location information requests (from gateway MSC to HLR, for incoming calls);
- authentication data requests (from VLR to HLR/AuC).

Closely associated with each MSC (usually integrated into it) is a *visitor location register*. The VLR stores the current cell location and subscription information for every mobile actually in its MSC's coverage area. This means that the MSC can access this information readily, without having to wait for information requests to traverse the network to the HLR.

Each GSM network has (logically, at any rate) one *home location register*. An HLR stores every user's permanent subscription record, including service details such as international access privileges and call-forwarding status. When a user activates his or her mobile phone, the local MSC/VLR requests the necessary information from the HLR, and keeps it until the user leaves that MSC's area. When a call from outside the network arrives at a gateway MSC, the MSC consults the HLR to find out which MSC/VLR to send the call to. This means that the HLR is a single point of failure for the whole network; if the HLR stops, incoming calls cannot happen. Therefore, HLRs are either hosted on very expensive fault-tolerant hardware, or else distributed across a number of computers. In the latter case, however, there still needs to be some single logical point of access for the gateway MSCs to ask at, so the single point of failure cannot be wholly eliminated.

Security

Each GSM network has an *authentication centre* (AuC) associated with its HLR. As part of the registration process (when the MS is switched on, and also at intervals thereafter), the AuC sends the SIM card a random number, which the SIM encodes using a secret key (known also to the AuC) and returns. The AuC compares this with a locally generated copy before authorizing access. The same random number and subscriber key are used to generate the transmission-ciphering key.

Handover

When an MS moves between cell areas, it has to be handed over (or 'handed off') to the new cell. The MS monitors the signals from the neighbouring cells, and regularly passes the signal strength information on to its current BSC. Handover (which includes assignment of a new traffic channel when a call is active) is done locally by the BSC, except when transferring to a cell belonging to another BSC. Such handovers, and ones between MSC areas, are managed by the MSCs.

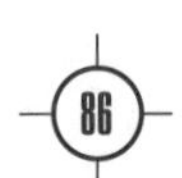

Caller tracing

In most countries, GSM operators (along with operators of other mobile network technologies) are obliged by statute to be able to locate a network user, to assist the police and the other emergency services. Locating users is also of interest to the network operators, to help them in diagnosing user problems, and to detect some kinds of fraudulent use.

In urban GSM networks, call tracing can be done to an accuracy of just a few metres.

In GSM networks, call tracing can be done to an accuracy of about 100 m in the country, where the cells are large, and down to just a few metres in cities. Call tracing is done by handing the MS over between a number of cells, and having each BTS report the round-trip delay time (which is part of the signalling system anyway). These delay times can then be used to triangulate to locate the user.

GSM standards

ETSI has published more than 100 standards relating to GSM, originating from the GSM Memorandum of Understanding (MoU) association. They are divided into families with coverage roughly as follows:

GSM 01:xx	General description, terminology and feasibility studies
GSM 02:xx	Definitions of services supported
GSM 03:xx	Network functions and architectures
GSM 04:xx	Air interface and related protocols
GSM 05:xx	Radio subsystem parameters
GSM 07:xx	Data interfaces for GSM terminals
GSM 08:xx	BSC/MSC interfaces
GSM 09:xx	Interworking with other networks
GSM 11:xx	SIM & SIM toolkit
GSM 12:xx	Management interfaces

GSM data services

GSM includes a range of data communications facilities.

Short message service

The point-to-point *short message service* (SMS) is a connectionless service, which enables users to send each other text messages. If the destination user is not reachable, the network's SMSC (SMS centre) will store the message and forward it later. Many SMSCs will also act as gateways to other access technologies, such as Web access. SMS also provides confirmation of delivery. However, SMS is severely limited by its maximum message size of 140 characters, and by its data rate of 9600 bps.

A recent improvement to SMS has been defined in ETSI standard 03.42, which proposes a data compression algorithm based on Huffman coding. This achieves a message length of around 200 characters. However, this modification has yet to be widely implemented.

Cell broadcast

Cell broadcast (defined in ETSI standard GSM 03.49) allows the network operator to send the same short message to all MSs within a given cell. The cell broadcast service is typically used to distribute location-related information, such as road traffic news. Up to 15 pages of 93 characters can be sent to any chosen cell, with selectable repetition rate.

Unstructured supplementary services data

Unstructured supplementary services data (USSD), like SMS, uses the shared signalling time slots of the GSM frame structure, rather than the allocated traffic channels. USSD is session-oriented and is immediate, not store-and-forward. Like SMS, its message size is limited (to 182 characters). USSD is often used as the bearer for WAP (*wireless application protocol*) over GSM networks.

Circuit switched data

Circuit switched data (CSD) uses a traffic channel to carry data at 9600 bps. As this is circuit switched, there is no limitation on message length. However, it means that a traffic channel has to be set up before data can be sent (and this can take as long as 30 seconds); also, it ties up a traffic channel continuously, regardless of whether or not there is data to send.

AMR

A recent addition to GSM has been the *adaptive multirate* (AMR) codec standard set (ETSI EN 301 704, 705, 707 & 712), which allows greater flexibility in allocation of channel capacity to voice calls. AMR supports both the carrying of several normal-grade calls in one time slot, and the use of multiple time slots to carry high-quality audio.

General packet radio service

The *general packet radio service* (GPRS) is a technology which offers high-speed mobile data services and IP switching, but which can be overlaid onto an existing GSM network without the need to change the radio interface equipment. It is therefore a very attractive option for GSM network operators which wish to offer advanced data services.

The GPRS data services are packet switched, offering a demand-sensitive variable data rate of up to 171.2 kbps burst rate. GPRS is particularly well suited to applications such as web access, where the connection time is very long, but no data is transmitted for much of the time. GPRS offers users a first sight of the high data rate network services promised by third-generation mobile networks, and is therefore sometimes called '2½ generation'. Figure 4.2 presents the major building blocks of a GPRS network overlay.

GPRS is particularly well suited to applications such as email access.

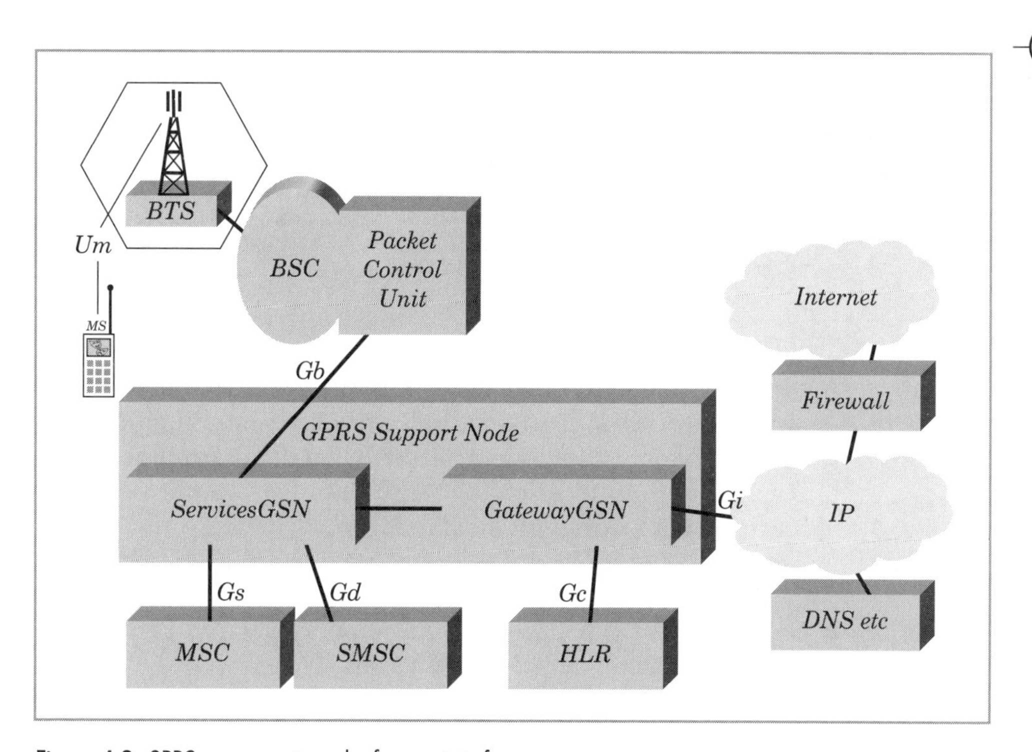

Figure 4.2 GPRS components and reference interfaces

Main GPRS technical standards	
GSM 02.60	GPRS service description Stage 1
GSM 03.60	GPRS service description Stage 2
GSM 03.64	GPRS overall description – radio interface
GSM 04.60	GPRS mobile station – base station system interface
GSM 04.64	GPRS logical link control layer
GSM 04.65	GPRS subnetwork dependent convergence protocol
GSM 07.60	GPRS mobile station
GSM 08.14	GPRS base station system – serving GPRS support node interface – Layer 1
GSM 08.16	GPRS base station system – serving GPRS support node interface – network service
GSM 08.18	GPRS base station system – serving GPRS support node – BSSGPRS protocol
GSM 09.60	GPRS tunnelling protocol
GSM 09.61	GPRS PLMN and packet data network interworking

MS

A GPRS mobile station has voice-handling features comparable to an ordinary GSM phone, but also has data-handling features which are likely to include a screen with some graphics capability, an alphanumeric keyboard, and a simple software applications environment such as a Web browser.

Three classes of MS are envisaged in the GPRS standards:

- class A: offers full simultaneous support of circuit switched voice and packet switched data traffic;
- class B: supports both voice and data calls, and responds to either without user pre-selection, but does not support both simultaneously;

- class C: supports both voice and data, but by manual pre-selection only.

Base station controller

A GPRS BSC is a modified version of a regular GSM BSC, with additional features to manage reservation of a controlled number of radio carrier time slots for packet traffic. Associated with each GPRS BSC is a *packet control unit* (PCU), which extracts the packet data from the radio time slots and passes it over a frame relay link to the nearest *serving GPRS support node* (SGSN).

GPRS support node

The *GPRS support node* (GSN) elements are the most significant addition to a GSM network to enable GPRS operation. A GSN consists of a *gateway GPRS support node* (GGSN) and a *serving GPRS support node*. These elements may be combined in a single GSN unit, or implemented as discrete items.

GGSN

A GSM network may have one or (more realistically) several *gateway GPRS support nodes*. The GGSN is the point of interconnect between the specialized world of GPRS and the standard IP network. The GGSN talks TCP/IP to one or more SGSNs, and connects not only to the external IP network but also potentially to GGSNs in other GPRS networks.

The GGSN links to its network's HLR (using SS7 MAP) to get routing information to set up the Layer 2 tunnelling for incoming data calls (tunnelling is described in Chapter 9). It also collects usage data and generates call detail records for analysis by accounting systems.

SGSN

There will also usually be multiple *serving GPRS support nodes* in a GPRS network. An SGSN talks frame relay to its associated BSCs, and controls the virtual circuits, called *packet data protocol contexts,* for the MS-GGSN connections. It also co-operates with its MSC (using SS7 BSSAP+) to co-ordinate voice and data services (as, for example, when a data Internet call carrying Web traffic to an Internet call centre needs to flip over to a voice call, without losing context).

The SGSN connects to the network's SMSC to enable SMS-GPRS service interworking. It also collects usage data and generates call detail records, like the GGSN.

Why GPRS?

GPRS is attractive to users because it makes wireless IP access available everywhere that GSM is deployed. It offers rapid call set-up and cleardown, with variable data rates of up to 171.2 kbps[5] (and up to 384 kbps if E-GPRS is deployed), and charging based on data volume, rather than on connect time.

For GSM network operators, GPRS offers a migration path to UMTS (*universal mobile telecommunication system*) without the immediate need to upgrade their radio equipment. The packetized radio interface uses the scarce radio spectrum efficiently, and similarly the packetized core network optimizes the utilization of fixed transmission infrastructure.

High-speed circuit switched data

A technically simpler alternative to GPRS is *high-speed circuit switched data* (HSCSD). This bundles several radio channels together to offer a (relatively) high-rate circuit switched service. If all eight channels on a GSM carrier are used, HSCSD can offer up to 64 kbps. HSCSD offers a *transparent*, fixed-capacity service, and also a *non-transparent* (NT) service, where one time slot is continuously allocated to the call, but others are added or removed by the network according to how much capacity is free.

The NT service makes good use of the available network capacity, without causing any more congestion than a normal GSM call. However, an HSCSD call cannot reduce its network requirement to less than one time slot, and so is not ideally suited to intermittent traffic such as Web access. For that reason, while HSCSD is easier to implement than GPRS, and in some networks has been implemented as a precursor to GPRS, it is unlikely to predominate in the long term.

[5] The exact peak data rate achieved depends on the equipment manufacturer.

CAMEL

The *customized applications for mobile enhanced logic* (CAMEL) features of GSM Phase 2 apply *intelligent networks* (IN) technology to provide value-added network services which can be supported when the user roams out of the home network, so long as the new network is also CAMEL-enabled.

EDGE

Enhanced data rates for global (or GSM) evolution (EDGE) technology is claimed to offer third-generation capacity on second-generation networks.[6] EDGE uses the same 200 kHz carriers as GSM, but by changing the modulation system manages to squeeze 48 kbps out of each channel, or 384 kbps out of each carrier. The cost of this is the need to replace the radio side equipment with new modulators, and in particular with more exactly linear amplifiers.

EDGE supports a GPRS variant called E-GPRS, mentioned above. Although EDGE is not fully standardized yet, its future appears to be assured in at least some niches. EDGE is likely to be deployed in the US, as an enhancement to GSM 1900 and IS-136 for third-generation networks. In Europe, its role may be restricted to being a fallback for any GSM network operators which fail to secure UMTS licences.

Although EDGE is not fully standardized yet, its future appears to be assured in at least some niches.

GSM variants

There are three other second-generation cellular systems which, while they are covered by the GSM standards, operate in different frequency bands and so are incompatible with ordinary 900 MHz GSM MSs. However, multi-band MSs can work with all of them. They are:

- *GSM 1800* (operating around 1800 MHz), alias PCS (*personal communications system*), alias DCS 1800 (*digital communication system at 1800 MHz*), alias E-Netz;
- *GSM 1900* (around 1900 MHz), alias *PCS 1900*;

[6] By manufacturers with more enthusiasm than exactness.

- *GSM 400* (developed to support the migration of NMT 400 systems to digital).

Other second-generation systems

GSM is far from being the only second-generation mobile system, although it is certainly the most widespread and the most rapidly growing.[7] The following are the main alternatives.

'CDMA'

Although CDMA is properly a generic name for a whole bunch of wireless transmission solutions, it is also used as a label for one in particular: cdmaOne™, which is defined by ANSI IS-95.

ANSI IS-95 defines a compatibility[8] standard for wideband mobile services, using CDMA spread-spectrum transmission. It includes both the radio interface (800 MHz) and the call-handing model, covering the processes for channel separation, power control, call processing, handoff and registration. IS-95 offers a data rate of 16 kbps; IS-95b extends this by bundling together four 16 kbps channels to offer 64 kbps, and IS-95c offers a packet-mode service with a peak data rate of 144 kbps.

Also, ANSI-J-STD-008 defines a standard for 1800–2000 MHz CDMA PCS systems. Interworking between ANSI J-STD-008 and IS-95A is addressed by standard TSB-74.

One of the advantages of using CDMA in a mobile application is its resilience to multipath fading, which affects FDMA solutions. Also, CDMA cellular has the advantage that as all cells can use the same frequencies, cell planning is made simpler.

'TDMA'

Similarly, the 'TDMA' label is used over-specifically to refer to the *digital AMPS* (D-AMPS) technology. AMPS is a widespread first-generation analogue system operating around 860 MHz. AMPS divides its spectrum into 30 kHz carriers, each carrying three or six voice calls (depending on the speech coding used).

D-AMPS adds a TDMA layer below the FDMA structure of the system that AMPS, rather like the system that GSM uses.

[7] As of mid-1999, GSM systems accounted for about 60 per cent of mobile subscribers worldwide.

[8] It does not concern itself, for example, with other issues such as service quality or reliability.

This approach means that D-AMPS MS can be made backward compatible with AMPS networks, which the Federal Communications Commission (FCC) insisted on for digital cellular systems in the US.

D-AMPS relies on standards ANSI IS-136 (formerly TIA[9]/EIA-136) and ANSI IS-54 (traffic on digital voice channels), and is promoted by the Universal Wireless Communications Consortium (UWCC).[10] UWCC and TIA have also developed a standard IN infrastructure for D-AMPS. Called *WIN*, it is based on ANSI IS-41 AIN, and on the ITU-T Q.1200 series IN recommendations. WIN works by applying to a mobile network the same sort of IN approach that is old hat in a fixed network: removing intelligence from the MSC and reducing it to an SSP, and so on. WIN offers services such as CLIP, call screening, messaging, 'one number' services, and private dialling plans, with a degree of compatibility between D-AMPS networks (comparable to the goals of the GSM CAMEL initiative). The primary standards for WIN are IS-41 (network architecture, interfaces and operation) and IS-53A (user services perspective).

PDC

Personal digital cellular (PDC or PDC-P, formerly *Japanese digital cellular*) is a TDMA technology local to Japan, defined by ARIB (Association of Radio Industries and Businesses, the Japanese wireless standards body). It offers voice services, and data services up to 16 kbps, using three-slot TDMA within an FDMA carrier spacing of 25 kHz. PDC occupies frequency bands around 800, 1500 and 1900 MHz. PDC is likely to be rapidly superseded, because of the expected early uptake of third-generation technology in Japan.

Third-generation (digital, data-oriented) cellular

IMT2000

The diversity of technologies and multiplicity of standards was widely held to have stunted the development of the market for second-generation mobile systems and services. Therefore, the

[9] Telecommunications Industry Association, a US industry gang.

[10] Another gang.

ITU-R led a massive exercise in international co-operation, supported by various R&D efforts worldwide, to unify the technologies emerging for the third generation. This programme was originally called the *future public land mobile telephony system* (FPLMTS), but then was re-badged as *international mobile telecommunications 2000* or IMT2000, not least because IMT2000 was considered to sound a good deal snappier than FPLMTS.

The data rate and mobility goals of IMT2000 were to offer:

- 2 Mbps for stationary users indoors;
- 384 kbps with 'pedestrian' mobility;
- 144 kbps with high-speed mobility.

These goals are compared with the achievements of GSM/GPRS (as a typical second-generation technology), and with those of fixed wireless systems such as LMDS, in Figure 4.3.

Other changes that are supported by IMT2000 are:

- integration with in-building and satellite wireless systems;
- migration from circuit-mode networks to mixed circuit and packet mode;
- expansion from symmetrical, real-time, delay-intolerant

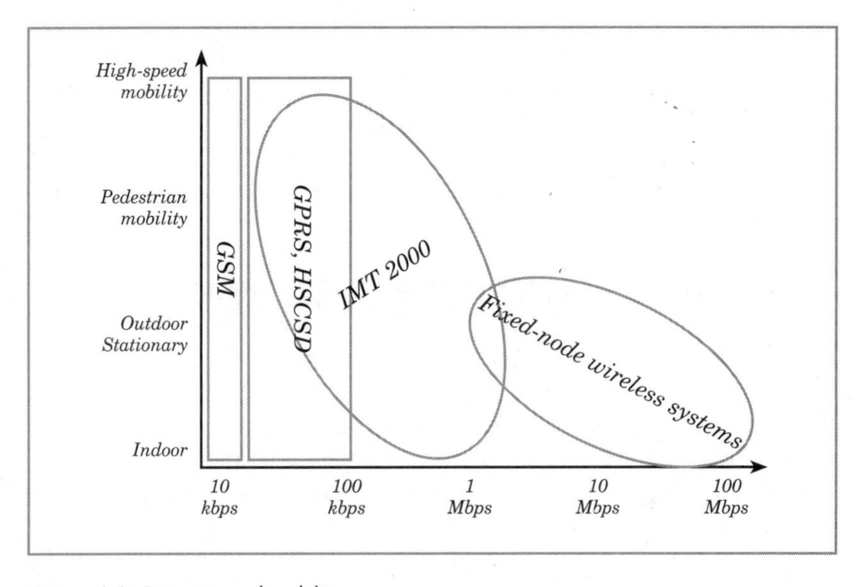

Figure 4.3 Data rate and mobility

voice services towards asymmetrical, delay-tolerant data and multimedia services;

- migration from telecoms-specific technologies to IP and conventional IT data networking.

Because of the weight of investment in diverse second-generation technologies, and the importance of offering second-generation network operators a migration path to the third generation, it has not been possible to standardize on a single technology for the third generation. There are therefore five third-generation mobile network technologies approved within IMT2000:

- IMT DS (*direct spread*), alias UMTS FDD or WCDMA;
- IMT MC (*multi carrier*), alias cdma2000;
- IMT TC (*time code*), alias UMTS TDD or TD-SCDMA;
- IMT SC (*single carrier*), alias UWC-136;
- IMT FT (*frequency time*), alias DECT.

Figure 4.4 illustrates their pedigree.

This diversity of air interfaces may limit roaming in third-generation networks, although some equipment manufacturers are proposing multi-way air interface chip sets.

The following sections outline the first four options; DECT is described later, in the context of its original role as a local area mobility technology.

ITU-R recommendations for IMT2000

M.816	Framework for services
M.817	Network architectures
M.818	Satellite operation
M.819	Adaptation to the needs of developing countries
M.1034	Requirements for the radio interface(s)
M.1035	Framework for the radio interfaces
M.1036	Spectrum considerations

UMTS

UMTS, the *universal mobile telecommunications system*, is an initiative led by '3GPP', a group of standards bodies including T1P1, ARIB, SMG (the Special Mobile Group) and ETSI. UMTS, although offered as a global standard, is likely to be particularly welcome in Europe, because it inherits a lot from GSM and GPRS.

UMTS, although offered as a global standard, is likely to be particularly welcome in Europe.

In Europe, 250 MHz of bandwidth has been allocated to UMTS, around 2 GHz. Licences are currently being allocated for UMTS operators, and commercial service availability is

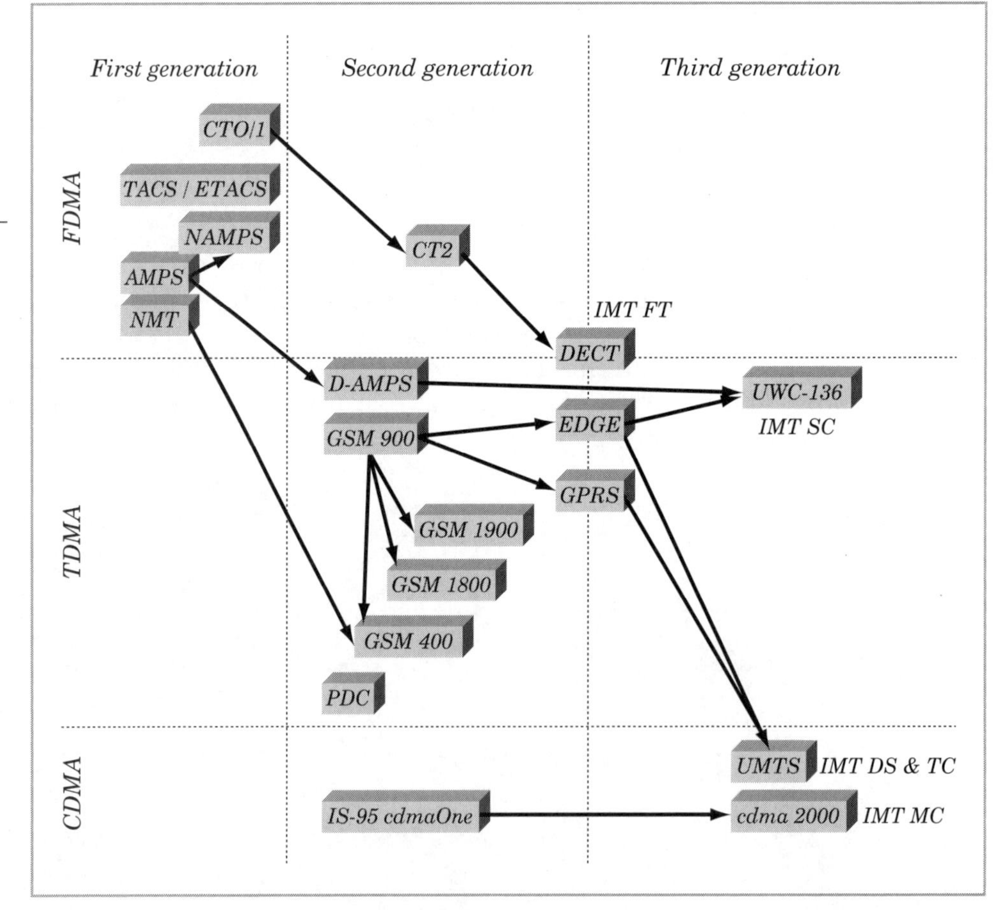

Figure 4.4 Relationships between the major cellular mobile systems

expected around 2002. The main components of a UMTS network are shown in Figure 4.5.

UTRA and UTRAN

The primary wireless access network technology of UMTS is called *UMTS terrestrial radio access* (UTRA), and its implementation is called *UMTS terrestrial radio access network* (UTRAN). As well as its mobile application, UTRA is relevant to fixed networks, where it may be adopted as a solution for in-building wireless connectivity.

UTRA uses *direct-sequence wideband CDMA* (WCDMA), with a 5 MHz bandwidth chipped at 3.84 Mcps. As with other CDMA mobile technologies, all the available bandwidth can be used in all cells. Inter-cell interference is managed through the inherent interference rejection of CDMA, and through very careful management of transmitter power levels. Because this arrangement allows an MS to communicate with two or more base stations simultaneously, it enables soft (make-before-break) handovers between cells.

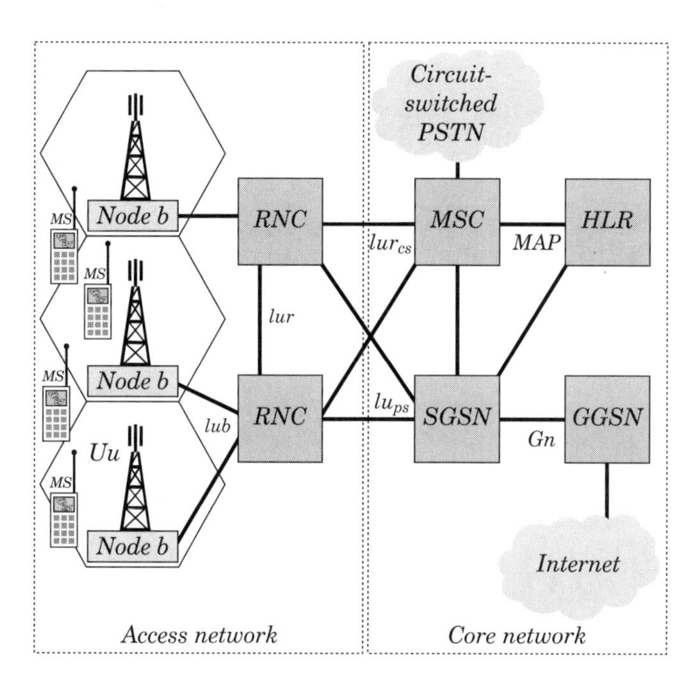

Figure 4.5 UMTS network elements and reference interfaces (Phase 1)

UTRA defines two options for duplexing (i.e. for accommodating both upstream and downstream channels). FDD mode (*frequency division duplexing*), alias FDCDMA, WCDMA or IMT DS, uses two separate 5 MHz bands for the two channels. TDD mode (*time division duplexing*) or TDCDMA (alias IMT TC), which is aligned with the Chinese TD-SCDMA, uses a single band.

Many of the components of Figure 4.5 will be familiar from GSM and GPRS network architectures (Figures 4.1 and 4.2). The UTRA *radio network controller* (RNC) takes the place of the GSM BSC, and the UTRA Node B takes the place of the GSM BTS, communicating with the RNC over ATM.

The UMTS standards are being released over a number of years, to lead the technology through a series of phases. In Phase 1, the core network is based on the GPRS and GSM core networks, running parallel packet-data and circuit-voice networks. In the longer term, UMTS will almost certainly evolve to an all-packet network architecture, as illustrated in Figure 4.6. However, the choice of packet mode infrastructure for future phases of UMTS (for example, ATM, IP or both) has yet to be finalized.

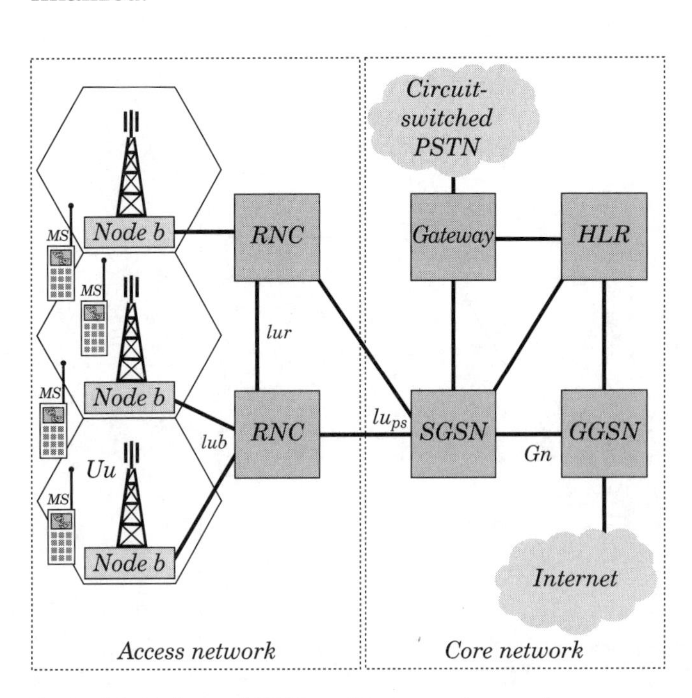

Figure 4.6 Evolution of UMTS to all-packet mode network

ETSI standards for UMTS

While many of the standards for UMTS are still in draft form, a few already have full ETSI standard status. Among these are:

- EG 201 718–1 Mobility management for evolved fixed networks
- EG 201 721 UMTS strategies
- EG 201 717 Virtual home environment (VHE) in the ISDN
- TR 101 111 Requirements for the UMTS terrestrial radio access system (UTRA)
- TR 101 146 UMTS terrestrial radio access (UTRA)

ERAN

UMTS also encompasses an alternative access network technology, EDGE radio access network (ERAN). ERAN replaces the UTRA Node b with an EDGE BTS, and uses an alternative form of RNC, for backward compatibility with EDGE networks.

cdma2000

cdma2000, alias IMT MC, is a third-generation technology promoted by 3GPP. The observant reader may note that 3GPP has already appeared as the promoter of UMTS. However, all will be made clear; cdma2000 is promoted by '3GPP *part 2*'. 3GPP's point of view is that just as the GSM community needs UMTS as a feasible migration path, so the IS-95 cdmaOne™ community needs cdma2000.

cdma2000, then, is based on ANSI IS-95 CDMA. Its air interfaces include a 3.68 Mcps direct spreading option and a multicarrier FDD mode. The 'Phase 1', '3G 1X' or 'MC1X' air interface (defined in TIA IS-200 1XRTT[11]) uses a 1.25 MHz band, and offers channels of 144 kbps. It also offers backward compatibility with IS-95 equipment.

The 'Phase 2', '3G 3X' or 'MC3X' air interface (defined in

[11] RTT stands for *radio transmission technology.*

TIA IS-200–A 3XRTT) offers a 2 Mbps service, either across 3×1.25 MHz, or over a single 3.75 MHz band. Also, in the short term, there is a migration option 'IS-95B', offering 64 kbps. Signalling in cdma2000 networks, as in IS-95, is based on ANSI IS-41 SS7.

UWC-136

The last member of the IMT2000 family for us to mention is UWC-136, alias IMT SC. Promoted by UWCC (the promoter of IS-136 TDMA), UWC-136 applies modified versions of EDGE and E-GPRS to TDMA networks. The air interface standard is ANSI IS-136HS, which is an extension of IS-136.

Fourth-generation terrestrial mobile systems

Work is already under way to define fourth-generation mobile systems.

Work is already under way to define fourth-generation mobile systems offering multi-megabit data rates. The ITU's fourth-generation programme is called the *mobile broadband system* (MBS).

Satellite mobile systems

Satellite systems are classified according to the altitude of the satellites:

- *geostationary earth orbit* (GEO) satellites, at around 36 000 km;
- *medium earth orbit* (MEO) satellites, at between 10 000 and 15 000 km;[12]
- *low earth orbit* (LEO) satellites, at between 500 and 10 000 km.

GEO mobile systems

Twentieth-century satellite mobile communications systems were almost exclusively GEO, and dominated by the International

[12] The gap above 15 000 km is to avoid the electrically difficult Van Allen belts.

Maritime Satellite Organization (Inmarsat). Inmarsat established a number of systems (Inmarsat-A, -B, -C, -M and -Aero), offering various mixes of circuit switched voice, circuit switched data (up to 64 kbps), packet switched data (2.4 kbps) and an SMS-like service at 600 bps. All of these systems required substantial terminals, ranging from a laptop device (with separate antenna) up to a lorry-top system with a 150-kg steerable antenna. The size of these terminals, the cost of the services, and the relatively low traffic capacity, meant that these services were unsuitable for the mass consumer market. The latest Inmarsat system, called Horizons, will offer data rates up to 144 kbps, and will operate in the 2 GHz band, in alignment with IMT2000. Still, the service will have restricted total capacity, and is targeted at specialized, well-funded users, rather than the general public.

MEO and LEO mobile systems

There are a number of ventures under way to provide *global mobile personal communications by satellite* (GMPCS) to the mass markets. The systems use between 10 and 66 satellites each, in MEO or LEO orbits. The transmission systems used are TDMA and CDMA, and the user data rates offered are modest, around 2.4–4.8 kbps. Even though the large number of satellites increases the capacity of the systems, none of them offers more than 150 000 erlangs, so they can never become the primary system for consumer mobile communications. However, the user terminals are almost as small as a terrestrial mobile phone, and the low orbits mean that the vexing 250 ms round-trip delay associated with GEO systems is not incurred.

Digital cellular PAMR

PAMR also has a digital cellular technology comparable to second-generation systems, called *trunked mobile radio*. This odd-sounding name is justified, but only just, as follows. The radio spectrum is divided up into many channels. The use of these channels is shared, subject to negotiation over a signalling channel. So, the bearer channels are available to all callers, in the same way that telephone trunks are.

There are a number of digital cellular PAMR networks, used mostly by the emergency services but also by motoring organizations and haulage fleets.

The dominant European digital cellular PAMR system is called TETRA (*trans-European trunked radio access*, or *terrestrial trunked radio*). TETRA originates from the TETRA MoU organization, and is standardized through ETSI. TETRA operates in various bits of spectrum between 380 MHz and 930 MHz. Like GSM, TETRA uses FDMA to divide the available bandwidth into carriers spaced at 25 kHz, providing 36 kbps each. TETRA then uses TDMA to deliver four channels within each carrier.

TETRA voice calls are encoded into 7.2 kbps.

ETSI standards for Tetra

There are more than 100 ETSI standards addressing TETRA, including:

EN 301 435: attachment requirements for TETRA terminal equipment

ETS 300 392 series: voice plus data (V+D) operation

ETS 300 393 series: packet data optimized (PDO) mode

ETS 300 395 series: speech codec for full-rate traffic channel

ETS 300 396 series: technical requirements for direct mode operation (DMO)

ES 201 658–660: digital advanced wireless service (DAWS)

TETRA offers normal point-to-point calling, as well as point-multipoint group calling. It can operate both via the network infrastructure and in *direct mode operation* between terminals (for example where fire fighters are inside a building and out of contact with the network). Also, mobile terminals can act as relays (for example, the fire engine may act as a relay point for its crew).

TETRA offers a range of data services:

- a *status* service, which delivers very short (16-bit) messages very quickly, even when there is a circuit call on the same terminal;
- a low data rate store-and-forward service akin to SMS, which is called *short data service* (SDS). It carries messages of up to 126 bytes, and offers a group messaging option as well as unicast;
- a 28.8 kbps *circuit switched service*, which was available before GSM HSCSD or GPRS, and thus secured an early slice of the mobile data market for TETRA;
- an IP packet data service based on the *assigned secondary control channel* (ASCCH).
- the next generation of TETRA systems will implement an extended *packet data optimized mode*.

Customer premises wireless technologies

While customer premises networks are largely outside this book's scope, some customer premises wireless technologies are so closely linked with telecommunications technologies that the two are merging.

Telephony-oriented technologies

In-building mobile systems standards began with analogue systems such as CT1 (renamed CT0). Now most in-building systems use either microcellular implementations of GSM, or digital cordless systems. The first European digital cordless standard was CT2, which was also used for some public access *telepoint* networks.[13] CT2 offers 40 voice channels per cell.

DECT is the most widely used cordless telephony standard in the world.

CT2 (and a short-lived CT3) has been superseded by DECT (*digital enhanced*[14] *cordless telephony*, alias IMT FT). DECT is standardized through ETSI (for a technical overview, see ETSI

[13] Telepoint means having a mobile unit with a very short range, and having base stations dotted around public places, usually deliberately visible. To make a call, the user has to stand near to a base station. Telepoint networks have not been widely successful because second-generation wide-area mobile technologies such as GSM and D-AMPS are so much more convenient.

[14] Formerly European.

EETR 178). It has been adopted in many countries outside the EU, and is the most widely used cordless telephony standard in the world. DECT uses a band between 1880 and 1900 MHz; within that, FDMA is used to provide 10 channels. Within each channel, TDMA and TDD are used to carry 12 duplex speech or 24 simplex data calls. DECT has a range of 10–30 metres indoors, and up to 100 metres outdoors. Multiple DECT cells may be operating in the same area, on the same frequencies. There is an automatic algorithm for channel selection, which minimizes the effect of interference.

DECT systems can be used as single-user in-home mobility solutions, or as the access medium for wireless PABXs, or even for fixed-terminal WLL. Dual-mode DECT/GSM phones are available, which will use DECT within the office building (thus connecting to the fixed PSTN), and switch over to GSM operation outdoors.

In Japan, there has been a singularly successful telepoint system, the *personal handyphone system* (PHS), originating from the Nippon Telegraph & Telephone Corp. (NTT). The US equivalent is the *personal access communications system* (PACS). Both of these systems share many features with CT2 and DECT.

Wireless LANs

There are also a number of overlapping and competing wireless LAN technologies, as follows.

Infrared

The Infrared Data Association (IrDA) offers a technology based on free-air optical communication. The IrDA system is widely implemented in laptops, personal assistants and mobile phones. In many ways it is very limited: its range is only about 1 metre, it is directional to within a 30 degree cone, and so is only really practical for stationary use. Although it offers data rates from 9600 bps to 4 Mbps,[15] there is only a point-to-point mode (no real LAN emulation), and there is no encryption. However, IrDA entered the wireless LAN market early, and so has secured a large slice of the market.

Infrared transmission is also supported among the many media options of IEEE 802.11, offering rates varying between 1 and 10 Mbps according to the modulation scheme used.

[15] A 16 Mbps version is under development.

RF 2.4 GHz band

Several technologies have converged on an unregulated RF band around 2.4 GHz, including:

- *Bluetooth*, which offers a 721 kbps encrypted point-multipoint service, over a range of about 10 metres, using a hybrid direct sequence and frequency-hopping spread spectrum radio interface. It has a capacity of three voice channels and seven data channels per piconet;
- the Wireless LAN Alliance (WLANA)'s system, defined in IEEE 802.11, which offers a point-multipoint LAN emulation running at 1–2 Mbps, with two alternative frequency-hopping and direct-sequence spread spectrum systems.

HIPERLAN

ETSI is standardizing a system called HIPERLAN (*high-performance radio access local area network*).[16] Hiperlan operates in two frequency bands, one around 5 GHz and another around 17 GHz. HIPERLAN/1 offers a wireless ethernet, while HIPERLAN/2 offers ATM and IP services. HIPERLAN has an operating range of about 200 metres, and can offer data rates of 24 Mbps (at 5 GHz) and as much as 155 Mbps at 17 GHz.

Mobility-related developments

Soft phones

The mobile terminal is becoming an increasingly open environment for executing software applications, which themselves may be downloaded over the radio interface. ETSI SMG 4 has defined a SIM toolkit, which is a software environment for applications executing in a SIM card. As mobile phones converge with personal organizers, other operating systems for handheld devices are MexE (the *mobile station application execution environment*) and Symbian EPOC.

Java has become the normal programming language for these devices, and there is a specific Java dialect, JavaCard, for use in the very small computing environment of a SIM card.

[16] For starters, see ETSI ETR 069 (HIPERLAN services and facilities) and ETSI ETR 133 (HIPERLAN system definition).

The range of possible applications for programmable mobile data communications devices is growing rapidly, and includes:

- personal banking;
- field force dispatch;
- electronic purchasing;
- vehicle tracking;
- email access;
- web browsing;
- unified messaging;
- prepayment for network services;
- local information services (based on cell identifier, signal strengths, signal delays or an integral GPS);
- directory enquiries.

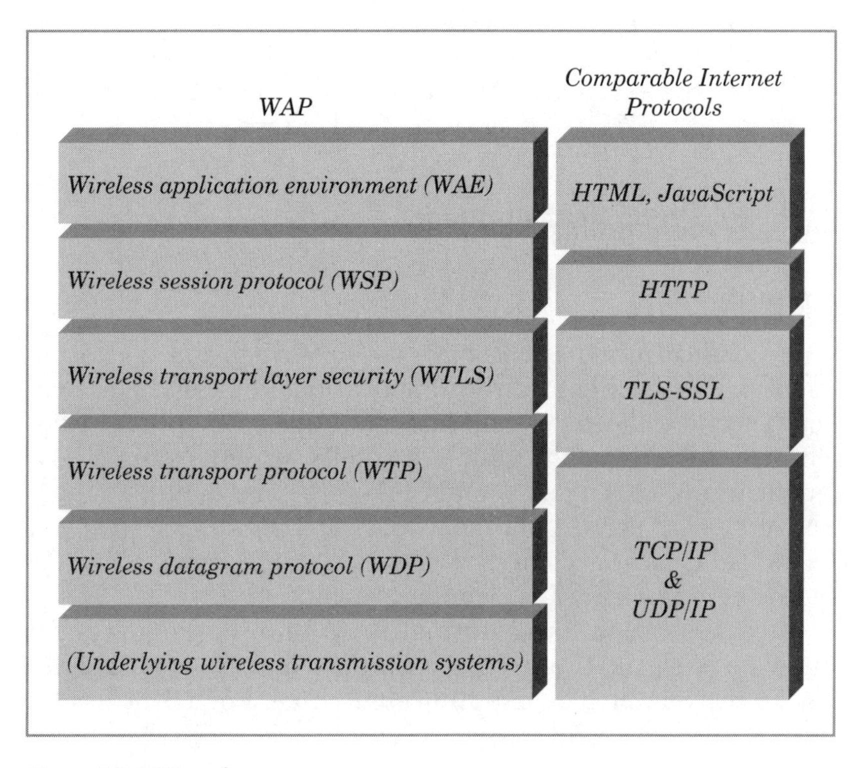

Figure 4.7 WAP stack

WAP

The so-called *wireless application protocol* (WAP, even, alas, W@P) is actually a whole set of protocols, aimed at providing a uniform communications environment across a wide range of wireless networks. WAP originates from the WAP Forum (a gang of more than 120 equipment manufacturers), and the WAP specifications (currently WAP version 1.2) are available from them at http://www.wapforum.org/.

The so-called 'wireless application protocol' is actually a whole set of protocols.

The WAP protocol stack (shown in Figure 4.7) is designed to reduce the computational load on the mobile device, making it a thin client, with as much of the work as possible pushed back into the network and the application servers.

The stack supports a huge range of underlying network technologies including GSM 900 (SMS and USSD), GSM 1800, GSM 1900, GPRS, PDC, IS-95 CDMA, IS-36 TDMA, TETRA, DECT and PHS (all of which are discussed elsewhere in this chapter). The WDP adapts the particular network technology to provide a uniform, network-independent transport service similar to ordinary UDP. WTLS adds privacy, data integrity and authentication services. WTP offers a range of remote transaction invocation services.

The WAE includes:

- WML, the *wireless markup language* (which replaces an earlier HDML – *handheld device markup language*), is an XML derivative which does a job comparable to HTML but with a view to execution in the environment of a 'microbrowser' in a handheld device with a small screen and limited computing power;
- WMLScript provides scripting facilities comparable to JavaScript, for the microbrowser environment.

The WAE may also soon be extended to include a programmers' API (the *wireless telephony application* – WTA), convergent with the SIM toolkit.

As Figure 4.8 shows, a mobile device can access the Internet through a WAP proxy, which converts between IP and WAP network and transport protocols, and which condenses the content of HTML Web pages to fit within the constraints of WML for onward transmission to the mobile device.

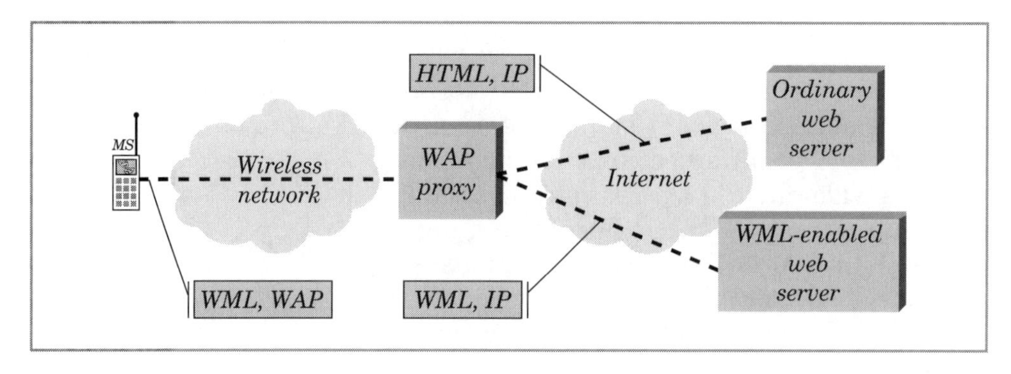

Figure 4.8 WAP and the Internet

While WAP enjoys immense support from industry, because standardization was essential for developing the wireless data market, it is not without its limitations:

- its encryption facility (in WTLS) extends only to the WAP proxy, and not all the way to the HTML server, so a continuous secure environment is difficult to achieve;
- its data compression capability is limited;
- although WSP includes a *push* facility, this is not supported by corresponding facilities on the internet stacks used by the fixed networks to which WSP has to connect.

Further reading

For a good, more extensive, review of mobile access network technologies, see Nilsson (1999). If you like detail, and want to get really into the RF interfaces, look at Wesel(1997), GSM in particular is reviewed concisely and clearly in Peersman (2000), which also gives a lot of detail about SMS.

5 Circuit mode multiplexing

Multiplexing means using a single-bit transport to carry more than one stream of traffic at a time. This is desirable for commercial reasons. For example, where the traffic streams are slow (such as a voice call at 64 kbps) but the bit transport is fast (say, a 2.4 Gbps fibre optic line system), to use one bit transport per traffic stream would be grossly wasteful. By using their bit transports efficiently, network operators can improve their profitability. Looking at it another way, if we expect to have to carry 1000 simultaneous calls between London and Manchester, then if we can get away with just one or two bit transports, rather than 1000, we have made our lives easier.

There are all sorts of ways of multiplexing digital signals, but I shall ignore most of them here. Frequency division multiplexing (alias wavelength division multiplexing), TDMA, CDMA and others are used mostly within the domain of the bit transport technologies reviewed in Chapter 3 and Chapter 4. Some of them are touched on in Chapter 2.

The two multiplexing technologies of most interest to software engineers are called PDH and SDH. They have a lot in common. Both are forms of *time division multiplexing*. The idea of TDM is to impose a regularly repeating frame structure on the bearer bit transport, and then to allocate parts of this structure to the various signals to be carried. Figure 5.1 illustrates the idea.

There has to be a conspiracy between the kit at the two ends of the link, about the frame structure.

Clearly, to do this there has to be a conspiracy between the kit at the two ends of the link, about the frame structure. There also has to be an agreement about what the bit rate will be. This is far less simple than you might imagine. Suppose that the nominal rate is 2.048 Mbps. Suppose further that you are a device that is trying to interconnect two bit transports. The incoming transport is perhaps running a smidgen above the nominal rate because the clock at the other end is running slightly fast. The outgoing transport is running a smidgen below the nominal rate. You are receiving slightly more bits than you can get rid of. What do you do? Buffer them up for ever? You will have to dump some of them sooner or later.

Nobody will be pleased to lose a proportion of the bits that they are paying you to convey, so something has to be done about it. The first thing is to try to keep all the devices in your network running to the same clock speed. So any transmission network will have a synchronization overlay network. This will start from some marvellous caesium clock and try to propagate its time reference simultaneously to all points in the network. This limits the problem, but it isn't perfect. There is always a little drift between the nodes of the network.

That is why transmission networks, which we might like to be *synchronous* (same-clocked), are in fact *plesiochronous*[1] (that is, similar-clocked). Their different solutions to the problem of

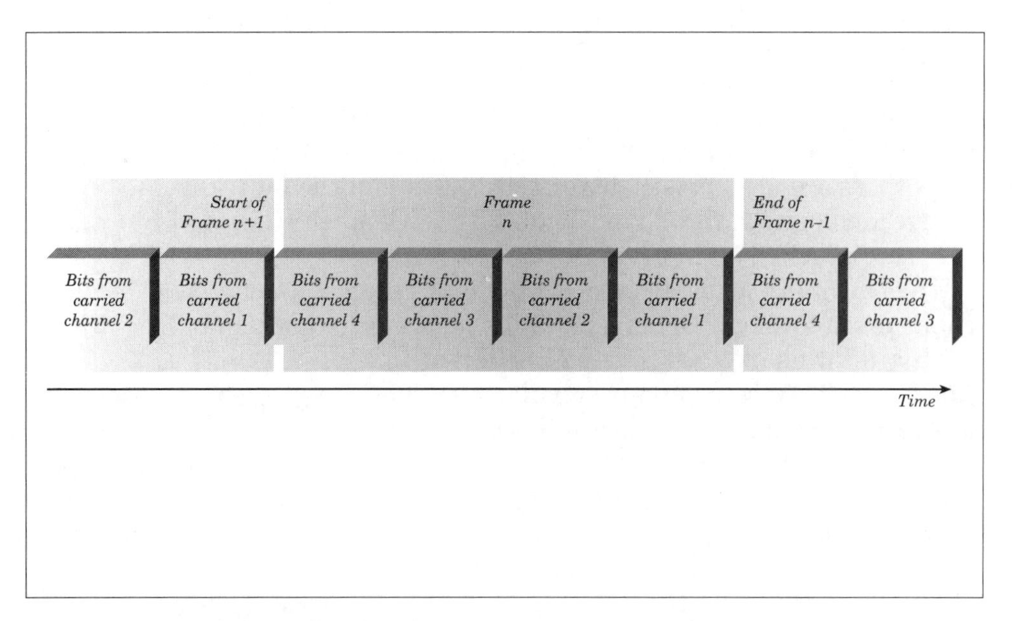

Figure 5.1 A (simplified) TDM frame sails by

[1] Please take care to pronounce that properly. To say 'plesio*syn*chronous' is incorrect as well as ugly.

clock drift are what best characterize and differentiate the PDH and SDH multiplexing systems.

PDH

PDH stands for the *plesiochronous digital hierarchy.* It is called a hierarchy because there are a set of defined transmission bearer rates, where a high rate bearer carries a number of data streams from the next rate down. Each of those streams may itself be a bearer carrying some lower-rate data streams, and so on. PDH is a kind of TDM, as is SDH, and as are various other technologies. 'TDM' is often used over-exactly to mean PDH, so care must be taken to avoid confusion when 'TDM' is mentioned.

The standard rates defined by CEPT and the ITU-T for Europe are shown in Table 5.1, which also shows how the Americas use a very similar, but subtly different set of rates, and Japan uses a third set of rates. Thus the PDH transmission networks of the world are divided into three islands of rates and

European	N. American	Japanese
	1.544 Mbps (T-1 or DS-1)	1.544 Mbps
2.048 Mbps (E-1)		
	3.152 Mbps (DS-1C)	3.152 Mbps
	6.312 Mbps (T-2 or DS-2)	6.312 Mbps
8.448 Mbps (E-2)		
		32.064 Mbps
34.368 Mbps (E-3)		
	44.736 Mbps (T-3 or DS-3)	
		97.728 Mbps
139.264 Mbps (E-4)	139.264 Mbps (DS-4E)	
	274.176 Mbps (T-4 or DS-4))	
564.992 Mbps (E-5)		

Table 5.1 PDH rates

there are special difficulties in the international gateways where they meet. All the rates have alphanumeric designating codes, but in particular the European '2 Mbps' rate (actually 2.048 Mbps) is well known under the name 'E-1', and the 34 Mbps rate under 'E-3'. The US equivalent of E-1 is T-1, at 1.5 Mbps. An E-1 will carry thirty 64 kbps data streams; an 8 Mbps will carry 4×2 Mbps; an E-3 carries 4×8 Mbps, and so on in multiples of four.[2]

The acute observer will note that the numbers do not stack up right. 30×64 is 1920, not 2048, and so on. The higher rate always goes a little faster than the sum of the channels that it carries. The reason is that each rate has a *frame structure* which contains the carried channels (called *tributaries*), plus some bits to identify the start of the frame, plus some for rudimentary alarm reporting, plus some bits for *rate justification*. These last bits are dummies, which can be dumped by a multiplexer if a tributary rate overrun requires it.

ITU-T recommendations for PDH

G.702	Digital hierarchy bit rates
G.703	Physical/electrical characteristics of hierarchical digital interfaces
G.742	Second-order digital multiplex equipment operating at 8448 kbit/s and using positive justification
G.742	Second-order digital multiplex equipment operating at 8448 kbit/s and using positive justification
G.743	Second-order digital multiplex equipment operating at 6312 kbit/s and using positive justification
G.744	Second-order PCM multiplex equipment operating at 8448 kbit/s
G.747	Second-order digital multiplex equipment operating at 6312 kbit/s and multiplexing three tributaries at 2048 kbit/s

[2] The ITU standards go in regular multiples of four channels to one, at each level. The North American and Japanese standards have multiples including four, six and seven.

G.751 Digital multiplex equipments operating at the third-order bit rate of 34 368 kbit/s and the fourth-order bit rate of 139 264 kbit/s and using positive justification

G.752 Characteristics of digital multiplex equipments based on a second-order bit rate of 6312 kbit/s and using positive justification

G.755 Digital multiplex equipment operating at 139 264 kbit/s and multiplexing three tributaries at 44 736 kbit/s

G.955 Digital line systems based on the 1544 kbit/s and the 2048 kbit/s hierarchy on optical fibre cables

Where, say, an 8 Mbps circuit is carried as a tributary within a 34 Mbps bearer, the whole 8 Mbps is carried transparently. The higher rate bearer does not keep track of which bits in the 8 Mbps are user data and which are framing and justification. So the internal structure of a high-rate bearer can be painfully complex.

Up to a point, this works fine. Figure 5.2 shows a circuit (perhaps a leased line) crossing a transmission network. At the service delivery points, the circuit appears as, say, an E-1, 2 Mbps. As it gets further into the network, it is multiplexed with more and more others, at rates of 8, 34 and finally 140 Mbps, to cross the network, and then it is demultiplexed at the other side.

PDH *multiplexers* are, from the software point of view, simple. There is rarely anything to configure in them; they just have four tributary ports at the lower rate, an *aggregate* port at the higher rate, and a *synchronization input*. Their ability to

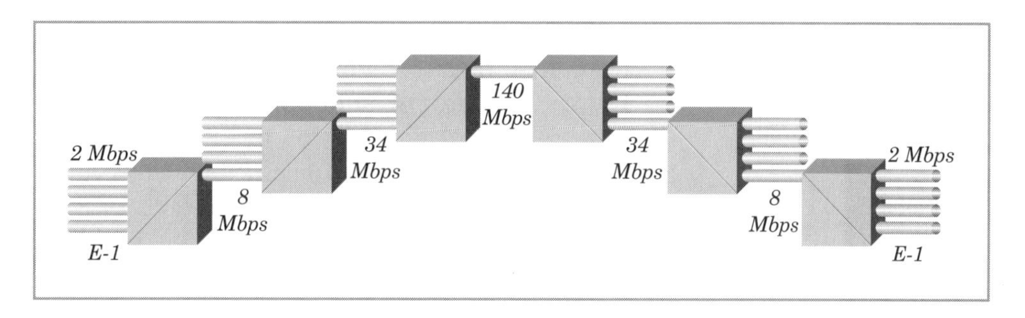

Figure 5.2 PDH multiplexing

raise network management alarms is limited; they can produce a signal called AI (*alarm indication*) whenever they detect a problem. One of the problems of transmission network management is that if, say, the highest rate bearer in Figure 5.2 were to be broken (perhaps by a careless road excavator), all the 256 E-1 end points would report AI. Such a flood of alarms is, for the network manager, tiresome.

PDH is unsatisfactory. If it were not, SDH would not have been necessary. The major problem is as follows. Boggis & Co wants a 2 Mbps E-1 connection to somewhere. GlobelCo has a fibre running past Boggis' door, running PDH at 565 Mbps. It is only half used, so GlobelCo would love to offer 2m of it to Boggis. But the only way to do that is to have:

- a mux to get down to 4 × 140 Mbps;
- a mux to get one of those down to 34 Mbps;
- a mux to get one of those down to 8 Mbps;
- and a mux to get one of those down to 4 × 2 Mbps;
- and then 4 more muxes to reassemble the 140 Mbps!

This is called a *mux mountain*, and the case of Boggis is far from the worst that can arise. The way the justified bearers are all nested within each other makes doing it in a single box impractical. The expense of it is horrible just in muxes, not to mention the small building to house them and the diesel generator to power them.

Diversity/restoration overlays

If the network operator commits to providing a reliable service, some provision has to be made for continuing transmission when a link fails.

If the network operator commits to providing a reliable service, some provision has to be made for continuing transmission when a link fails. Figure 5.3 shows a simple example. Here, the normal circuit route runs over a microwave link. If that fails, the *diversity switches* automatically re-route the traffic over a *diverse route*. To reduce the number of single points of failure, the diversity switches have dual-redundant power supplies in this example.

Fractional E-1/T-1

The E-1 or (North American T-1) circuit is a convenient way of delivering bandwidth into a commercial customer's premises (the end of a circuit is sometimes called an 'appearance', or a 'G.703', after the ITU-T recommendation). But often a customer will want only to buy maybe 5 × 64 kbps. It would be inconvenient for the telcos to use anything less than a 2 Mbps G.703, so what happens is that a 2 Mbps channel is installed, and the customer only gets to use an agreed part of it. The unused channels are dumped in the network via a process called grooming, which is described in Chapter 7. An advantage of fractional E-1/T-1 is that if the customer later wants more capacity, it can be provided without a new circuit installation at the customer's premises.

This can provide rapid restoration of service after a single link failure. However, it will fail if the diverse route is also in trouble. Another problem is that it uses up a complete set of transmission equipment on the diverse route, which is not used for most of the time. A more resilient and economical alternative can be to provide a mesh-like network architecture, where each route can be protected by more than one fallback path, and where any piece of spare equipment is protecting more than one circuit. However, this kind of service protection network is complex to build and to manage, and is practical only in the high-rate core of a network.

Two common variants on the normal four tributary/one aggregate PDH mux are the *jump mux* and the *add-drop mux*

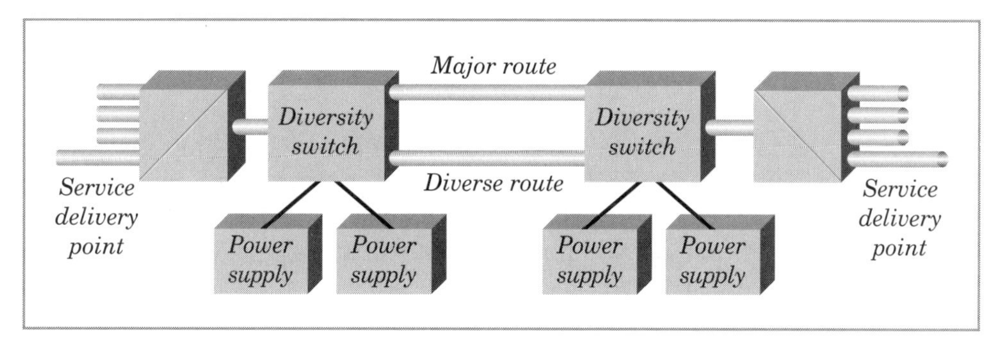

Figure 5.3 PDH diversity system

(ADM). The jump mux does more than one level of multiplexing in a single box; for example, from 1 * 34 Mbps to 16 * 2 Mbps.

The add-drop mux or *drop and insert mux* is effectively a pair of muxes connected back to back, with facilities for breaking out one or more tributaries, as shown in Figure 5.4.

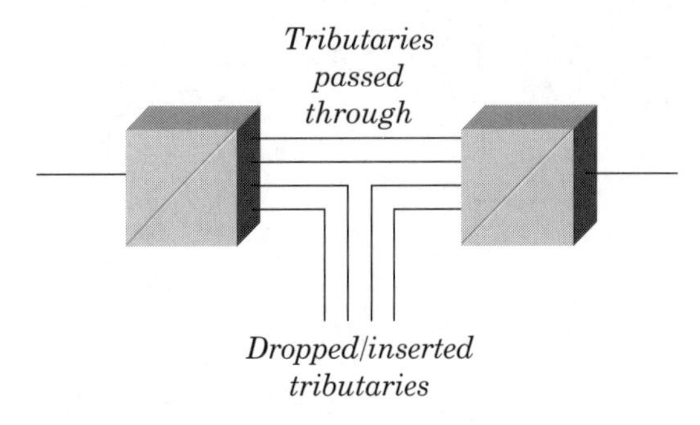

Figure 5.4 The add-drop mux

SDH

So, with PDH, path protection can be costly, and getting at unused capacity can be impractical. SDH, the *synchronous digital heirarchy*, also known as *Sonet* is an attempt to remedy these problems, and to offer *bandwidth on demand*[3]. SDH has been a fair success, and consequently new PDH installation has generally ceased.

Stream structure

SDH frames are in some ways simpler than PDH. The simplification is that, whatever rate they are at, there is only a single justification section. When a high-rate link carries some lower-rate bit streams, it does not (as PDH would) carry with them each one's justification bits. Figure 5.5 illustrates the difference.

Furthermore, the justification is done in byte chunks rather than in odd bits, so it is easier to handle with a computer. These differences from PDH mean that no matter how high a rate the SDH stream is running at, and no matter how many other streams are carried in it, it is comparatively easy to extract and add tributaries.

[3] Note that an SDH network is no more synchronous than a PDH one. SDH and PDH are both designed for imperfectly-synchronized mnetworks.

Selected ITU-T SDH recommendations

G.707	Network node interface for the synchronous digital hierarchy (SDH)
G.774	Synchronous digital hierarchy management information model for the network element view
G.780	Vocabulary of terms for synchronous digital hierarchy networks and equipment
G.783	Characteristics of SDH equipment functional blocks
G.784	SDH management
G.803	Architecture of transport networks based on the synchronous digital hierarchy
G.841	Types and characteristics of SDH network protection architectures

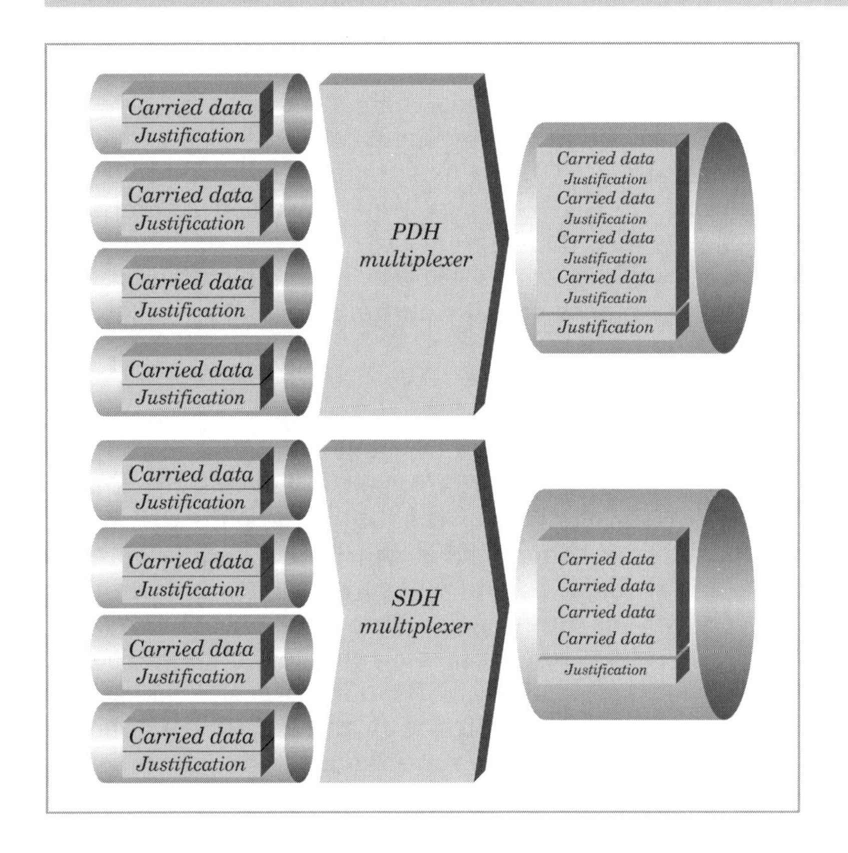

Figure 5.5 PDH and SDH justification structures

Another improvement on PDH is that the SDH standards include a lot of network management functionality, so that SDH equipment can be remotely controlled. Therefore, bandwidth on demand becomes a possibility.

SDH terminology distinguishes between the physical rate of a channel, and the effective rates of the traffic streams that it carries. There are actually two sets of standards, and two sets of physical rates, one called SDH, defined by ITU-T, and one defined by Telecordia, called Sonet (a contraction of *synchronous optical network*). ITU SDH physical channels are called *synchronous transport modules* (STMS), while Sonet ones are called *synchronous transport signals* (STS) or *optical channels* (OCS), as shown in Table 5.2.

An SDH simplex trail can meet specialized needs while saving on network resource.

The carried traffic within an SDH frame is called the *payload*. The payload can be one or several *virtual containers*. These come in standard sizes shown in Table 5.3.

A circuit through an SDH network is called a *trail*. Trails are characterized by obvious things such as their VC size and their end points, and also potentially by some refined features such as simplex[5] operation. PDH, in its simplicity, makes all circuits bi-directional. An SDH simplex trail can meet a number of needs (for example, distribution of broadcast services) while saving on network resource. Another example of SDH refinement is the capacity to provide multidrop trails.

Equipment

There are two types of SDH equipment of interest to the software engineer. Firstly, the *add-drop mux*, which is also known as the *drop and insert mux* or the *jump mux* (cf. PDH, where those are different things). This will have typically an SDH ring running through it, and will have tributary ports from which VCs can be multiplexed into and out of the ring. Secondly, where two rings intersect, the device for extracting VCs from one ring and inserting them into the other is called a cross-connect (q.v.). In Figures 5.6 and 5.7, the ADMs are shown as triangles and the cross-connects are shown as χ's, which is a common symbol for either a *cross-connect* or a circuit switch.

Physical channel rate	ITU SDH name	Sonet names
51.84 Mbps		STS-1, OC 1
155.52 Mbps	STM-1	STS-3, OC 3
466.56 Mbps		STS-9, OC 9
622.08 Mbps	STM-4	STS-12, OC 12
933.12 Mbps		STS-18, OC 18
1244.16 Mbps		STS-24, OC 24
1866.24 Mbps		STS-36, OC 36
2488.32 Mbps	STM-16	STS-48, OC 48
9953.28 Mbps	STM-64	STS-192, OC 192
39813.12 Mbps	STM-256	STS-768, OC 768

Table 5.2 SDH physical rates

VC–11[4]	1.5 Mbps
VC–12	2 Mbps
VC–3	34 Mbps
VC–v4	140Mbps

Table 5.3 SDH virtual containers

[4]Note that VC-11 and VC-12 are spoken as 'one-one' and 'one-two', *not* 'eleven' and 'twelve'.

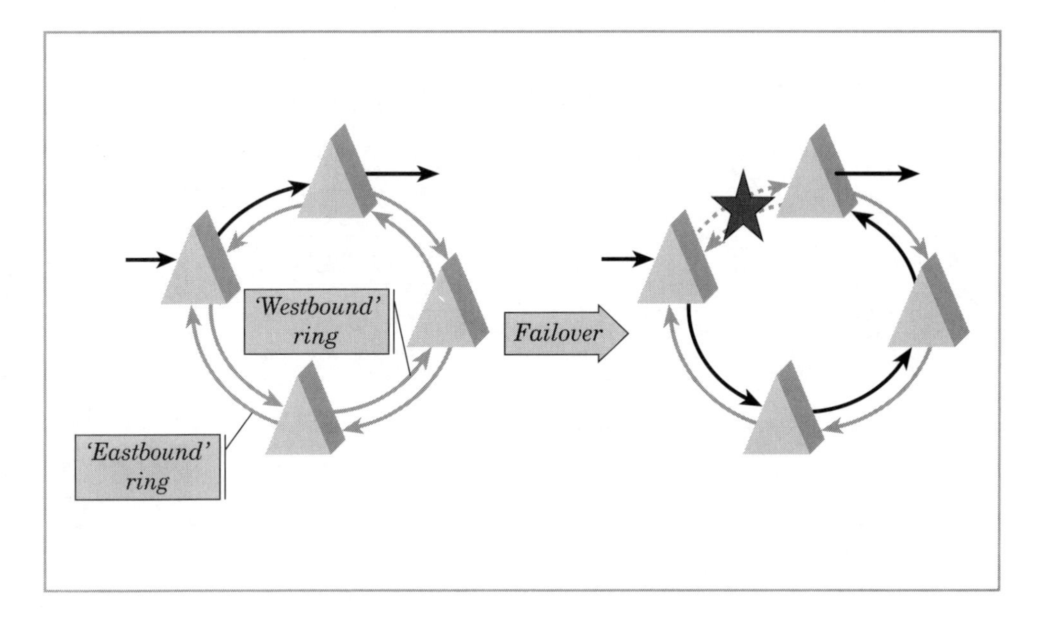

Figure 5.6 Shared protection ring

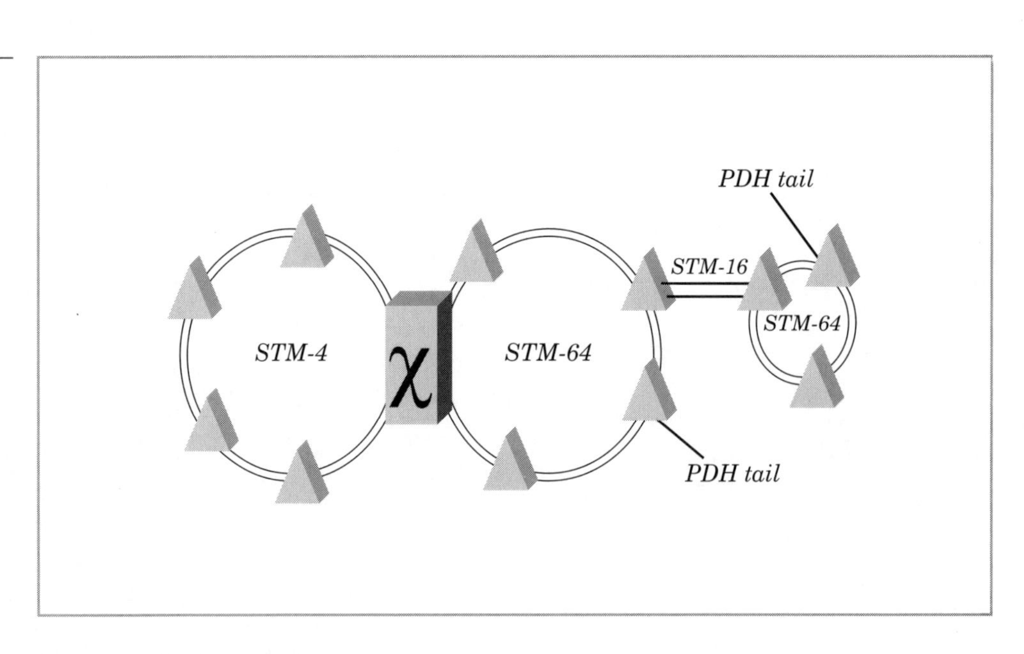

Figure 5.7 Multiring network

[4] I.e. unidirectional.

Network architectures

The archetypal SDH subnetwork is a bi-directional self-healing *shared protection ring* or SPRING, as shown in Figure 5.6. There are two rings of fibre running in opposite directions (called by convention 'east' and 'west', regardless of the geography). The idea is that if the ring is broken, the muxes on either side of the break can quickly loop back the broken ends, and send the traffic 'the wrong way' round the ring. This is in practice not always achievable, not least because of the enormous complexity it adds to the network management task of allocating capacity on the various segments of the ring.

While we are considering network management and pseudo-geography, note that another two common terms, originating in SDH but now used more widely, are 'north' (meaning towards a higher-level network management system) and 'south' (meaning away from the network management system, towards the network).

SDH networks usually comprise a number of intersecting rings. Often there is a high-speed national ring, with lower-speed rings round the edges, serving local areas. As Figure 5.7 shows, SDH does not have to be ring-shaped, and rings are often mixed with sections of more arbitrary topology. SDH networks also are rarely pure SDH; as end-user equipment rarely has SDH interfacing capability, the final leg to the customer premises is often PDH E-1/T-1. The provision of this access leg is sometimes curiously called 'swinging a tail'.

Management

The network management of SDH is complex. Because SDH boxes are so flexible, there are a large number of parameters inside them to control. And we have already seen how the provisioning of a trail across a single ring can incur quite complex considerations. Add to this the complexity of multiring architectures and the common desire to plan network usage ahead in time, and the complexity of SDH provisioning becomes challenging. Similarly, the surveillance of SDH networks is vastly more complex than that of their PDH counterparts.

Network management is also the key to SDH. If (as I suggest) SDH has yet to fulfil its promises of bandwidth on demand, it is because the network management systems have not got their act together yet. When a new SDH product is released, sometimes it is just the same old network element, but with a new network management system, which can unlock some more of its potential.

Further reading

For a good account of PDH and SDH/Sonet, see Flood, J.E. and Cochrane (1995).

6 Circuit mode signalling

Basic signalling events

'Signalling', in the context of a circuit switched network, means the exchange of requests and information to enable a network to establish, monitor, and clear down the calls which its users request. There are many signalling protocols in use, but they all have a lot in common, because they are all addressing essentially the same task. Figure 6.1 illustrates the signalling which goes on between the user terminals and the network, to set up and clear down a call. These signalling events occur, with small variations, for most (if not all) circuit switched services. Also,

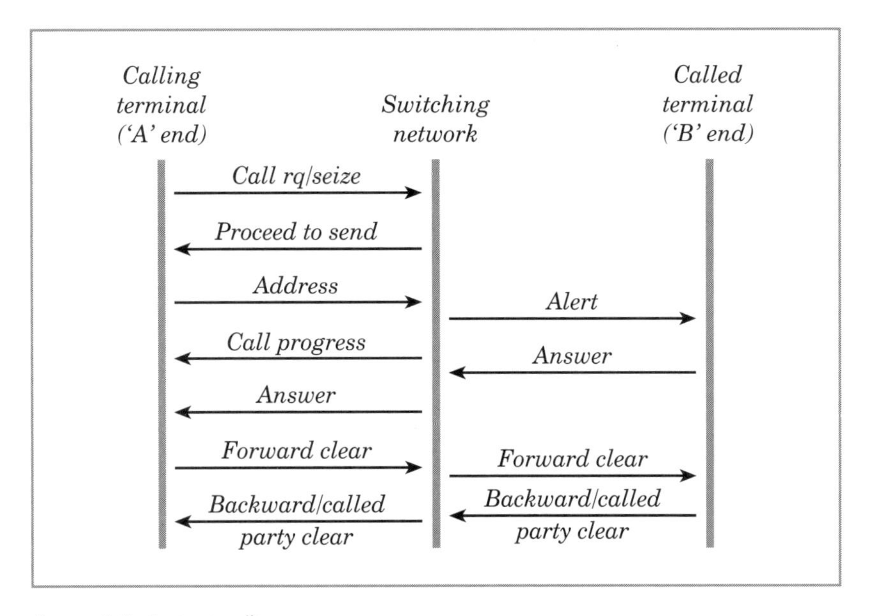

Figure 6.1 Basic signalling events

while Figure 6.1 illustrates what goes on between the users and the network, the interactions between the nodes within the network are remarkably similar.

Here is how Figure 6.1 applies to an ordinary voice call.

1. One network user (the 'calling party') wants to call another one (the 'called party'). Each of them has a terminal on the network; in this case, they have telephones. There is a common convention by which the calling party's end of the call is labelled the 'A end', and the called party's end is the 'B end'.
2. The calling party gets his or her terminal to send a *call request signal* to the network. In this case, he or she does this by going off-hook (picking up the telephone handset). Sometimes call request is also called *line seize*. This expression comes from situations where several user terminals have to share the same piece of network infrastructure (as in the case of a party line); seizing the line means establishing temporary exclusive access to it.

3. The network responds to the call request with a *proceed to send* signal, which is an indication that the network is ready to accept address information. In our example, this is done by the network producing a dial tone.
4. The calling party sends the *address* of the desired called party. In this case, it is their telephone number.
5. When the address has been signalled, the network validates it, establishes a route to the called terminal, and then sends an *alert* signal to the called terminal; that is to say, it makes the called telephone ring.
6. At various points in the call, the network may send call progress signals to either party. A good example of this is when the alert signal is sent to the called party, a signal called *alerting* or *alert indication* is sent to the caller. In the telephony example, this is the simulated ringing sound which the caller hears.
7. With luck, the called party will respond, and the B-end terminal will send an *answer* signal (by going off-hook) to the network. The network will propagate this to the A-end. In normal telephony, this appears to the user as

the end of alerting tone, rather than as a distinct positive signal. The network will then establish the speech path, and start metering the call for billing.

8 When the callers have finished their business, the call enters a cleardown phase. This can be initiated in two ways. Firstly, the calling party can request cleardown by going on-hook, in which case their terminal sends a *clear* signal to the network; the network stops metering the call, tears down the speech path, and sends a clear signal to the B end.

9 The second way is 'called party clearing', where the clear originates from the B end. Early telephony networks did not support called party clearing. However, this meant that a malicious user could call somebody and then fail to clear down the call, leaving the victim's equipment connected to the call indefinitely, and preventing other use of the victim's telephone. While this might be expensive to the malicious caller, it could be economically worthwhile in cases where the victim was, say, a competing service company. To address this, networks gradually introduced called party clearing. Now, most wireline networks support called-party clearing with a time-out, so that the called party can briefly put down their telephone without losing the call. On wireless networks, where radio bandwidth has to be conserved, immediate called party clearing is more normal.

A malicious user could call somebody and then fail to clear down the call, leaving the victim's equipment connected to the call indefinitely.

Two kinds of signalling

Signalling systems are divided into two kinds, according to the network architecture on which they rely. In *channel associated signalling* (CAS), the signal information travels down the same path as the users' call content (the *bearer* path). In *common channel signalling* (CCS), the signal information is sent along a network which is separate from the bearer network (see Figure 6.2). Typically, the signal paths for several terminals share a signalling channel, which is why it is called *common channel*.

Channel-associated signalling systems

Magneto calling and central battery

In the earliest days of telephony, electrical signalling was limited to the call request and alert signals, between end users and the human operators of manual telephone exchanges (or central offices). The signals were generated by hand-cranked magnetos, and it is from these that today's 48-volt AC ringing signal for wireline networks originated.

By the early years of the 20th century, a more convenient system known as *central battery signalling* (CBS) had been introduced. A battery at the local exchange applied a DC voltage across its end of the line.[1] At the user's end, the normal condition was an open circuit (so far as DC was concerned), with an

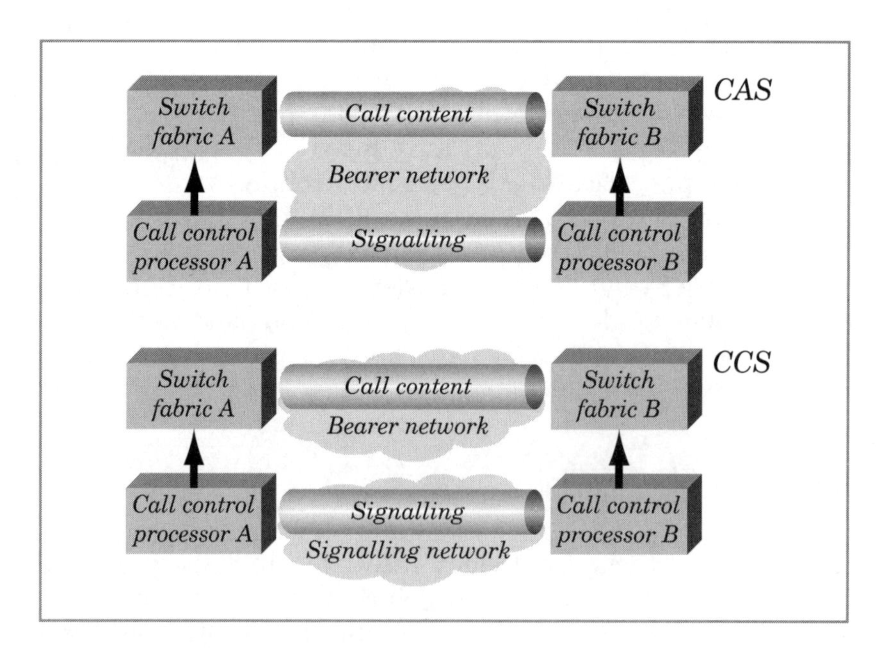

Figure 6.2 CAS and CCS

[1] Line voltages from 24 V to 60 V were used with various CBS systems, later becoming standardized at 50 v.

AC operated bell fed via a capacitor. When the phone was taken off-hook, a switch in the phone completed the DC circuit loop, causing a flow of current, which operated an indicator at the exchange. The exchange could call the customer (when on-hook) by applying an AC voltage to the line, which would pass through the capacitor and ring the bell.[2]

Very soon, the CBS system was developed so that the current supplied by the exchange battery also powered the carbon transmitter in the user's telephone, eliminating the need for a battery at each user station. Such systems were called *central battery* (CB) and were widely installed. Many CBS exchanges were also modified for CB working.

Loop-disconnect

When Strowger's automatic telephone switches were introduced, it was natural for them to adopt the CB signalling system, modified to permit the calling user to send address signals. This was achieved through the introduction of rotary-dial telephones. When the user picks up the telephone, the hook switch operates to complete a circuit loop back to the network (as in CBS or CB); that is how the call request is signalled. The rotary-dial mechanism then sends the address as a series of brief breaks in that circuit. The number '1' is typically signalled as a single break (or pulse); '2' is two, and so on up to ten pulses for '0'. The pulses are sent at the rate of 10 per second (approximately), and each pulse consists of a disconnection of the loop, and there are 10 digits available, so the system is called 10 *pulse per second* (PPS) *decadic loop disconnect*, or *loop-dis*.

Pulse encoding variants

Bizarrely, not all systems use the same encoding. In some cases, 0 is signalled as one pulse, up to ten pulses for nine. In at least one Scandinavian system, 0 is ten pulses, and then 1 is nine pulses, and so on up.

[2] Ringing voltages were usually about 70 volts at about 17 Hz.

Loop-dis signalling retained the AC alerting signal. A refinement was added in the case of party lines, where it is necessary to control which of the parties is alerted for a particular call. By putting series diodes in the users' bell circuits, and controlling the DC bias on the alerting signal, or by using 'one-leg-to-earth' ringing, the correct party's telephone can be made to ring.

Hookflash dialling

While loop-disconnect signalling has been largely overtaken by MF, most telephony switches still support it. On most telephones, therefore, it is possible, as a rather feeble party trick, to send the address by manually toggling the hook switch at roughly 10 pulses per second.

The 10 PPS decadic system was also extended to support signalling on amplified circuits between switches. As a DC path could not be guaranteed, the pulses were sent as tones, often at 2280 Hz, a frequency which was within the bandwidth of a copper-pair network, but which was particularly weak in human speech, and thus unlikely to suffer aliasing.

Multi-frequency signalling

The next development in user-network signalling was to replace the trains of DC pulses with single pulses of varying frequency tones, using the tones to encode the digits. Typically, *multi-frequency* (MF) signalling allows each digit to be sent as one 100-millisecond pulse, which compares well with the long pulse trains of loop-dis. A fair number of MF signalling systems were developed, but only a few of them persist.

In the access network, MF user-network signalling enabled the introduction of push-button 'touch-tone' telephones (at a time when the logic to turn a button push into a pulse train would have been costly) and offered faster signalling than loop-dis. For example, in the UK (and many other parts of the world), the MF4 *dual tone* MF (DTMF) system is used. In DTMF, each digit is sent as a combination of two tones, each of which is selected from a separate palette of four (so eight tones are used all told), as shown in Table 6.1. Of the 16 possible combinations, 12 are used in public networks, for digits 0–9, and two extra digits, '*' and '#', can be used to control advanced services. The remaining four combinations are sometimes used in specialized telephones designed to work with specific PABXs, for special-function keys.

In the trunk network, an MF system called R1, designed by Telecordia (and subsequently adopted by the ITU as *regional signalling system number 1*, recommendations Q.310–Q.332), was deployed extensively in the USA. R1 uses combinations of six frequencies to provide 12 distinct symbols (0–9, 'start' and 'end') in the forward direction, and another six for the backward direction. So unlike MF4, R1 was capable of simultaneous bi-directional working.

Less attractive (for the network owners) was the opportunity provided by R1 signalling for exploitation of the network by *phone phreaks*. Phone phreaks began in the 1970s to use MF tone generators (called blue boxes), applied to the mouthpieces of public telephones, to manipulate the routing of their calls in such a way as to get long-distance calls for local call charges (which in many cases were zero). At the time this abuse was difficult to detect or prosecute, the only possible charge being 'illegal abstraction of electricity'. While the general supersedure of R1 by SS7 has considerably reduced the opportunities, still the WWW is still rich in phone-phreaking lore.

Phone phreaks began in the 1970s to use MF tone generators to get long-distance calls for local call-charges.

The other significant trunk MF signalling system is ITU-T R2 (recommendations Q.400–Q.490), which is still used for some international signalling, and which has yet to be entirely superseded by SS7. R2 uses a shift system to increase the number of

Frequencies (Hz)	1209	1336	1477	1633
697	1	2	3	'A'
770	4	5	6	'B'
852	7	8	9	'C'
941	*	0	#	'D'

Table 6.1 DTMF frequencies

available symbols to 15. Of these, some are standardized for international working, and others have been used by national variants (including the obsolete UK standard MF2).

Common channel signalling systems

Why move to CCS?

CCS, by allocating a separate channel for signalling, offers a number of advantages over CAS, including:

- the signalling channel can be optimized for data traffic, and thus offer a higher signalling data rate;
- signalling can continue during the call without disruption;
- there is no possibility of the bearer traffic emulating control signals (as in phreaking);
- the topology of the signalling network can be separated from the topology of the bearer network, which can reduce the quantity of signal repeater equipment needed;
- the segregation of the signalling information facilitates its processing in computer-controlled equipment (it is only very recently that computers have been fast enough to carry any significant volume of bearer traffic).

CCS was first widespread in the now obsolete ITU-T *signalling system number* 6 (SS6 or C6: recommendations Q.251–Q.300), which used 4800 bps modem links. Now, however, all the major CCS systems use more specialized and higher-rate data communications facilities.

DSS 1

CCS appears in the access network in the N-ISDN user-network interface (described in Chapter 3), which uses a single signalling channel (the D channel) to control its multiple bearer channels. The physical layer protocols for carrying the D-channel and the bearer channels are well standardized, in ITU-T recommendations I.420 and I.421.

There is more diversity in the D-channel signalling systems that run on top of these standards. The generally applicable standard, in principle, is *digital subscriber signalling* 1 (DSS 1), which is defined in ITU-T recommendations Q.850 and Q.920–Q.957. In case the reader is not already overwhelmed with recommendation numbers, it is necessary to explain that Q.930, Q.931 and Q.932 are duplicated by I.450, I.451 and I.452 respectively.

Q.921 defines the data link layer protocol LAPD (*link access protocol, D-channel*), which is one of the HDLC (*high-level data link control*) family of protocols, running *asynchronous balanced mode* (ABM). The main specializations in LAPD are to adapt it to multidrop arrangements, where a number of user terminals are connected to a single network termination point and an extended window size to address long link delays (for example, over satellite links).

The message set for DSS 1 is documented in Q.931, and it covers the set of signals shown in Figure 6.1, with just a few extras.

Other ISDN protocols

DSS 1 is not, however, universal in real life. Because DSS 1 was late arriving, and because it contained more than the absolute minimum necessary facilities, a large number of bodies have defined their own alternatives, for example:

- equipment manufacturers Nortel and Ericsson have developed DMS100 ISDN and 5ESS ISDN respectively;
- for aviation use, Aeronautical Radio Inc has defined ARINC 746;
- in Australia, Austel has defined TS014;
- in France, there are VN 3 and VN4;
- in Germany, there is 1TR6 (essentially a profile of Q.931);
- in Japan, NTT has defined INS-Net;
- in Switzerland, Swiss Telecom has defined SwissNet;
- in the UK, BT defined *digital access signalling system* (DASS) and then a replacement, DASS 2, both of which are now obsolete;
- in the US, Telecordia has developed National ISDN 1 and National ISDN 2.

Within Europe, there is now convergence towards Euro ISDN, which is defined by ETSI (ETS 300 011, ETS 300 012, ETS 300 125 and ETS 300 102). Euro ISDN has also been adopted by operators in Israel, Malta and South Africa.

DPNSS and Q.Sig

Still outside the core telco network, CCS offers the possibility of connecting up a number of PABXs, within, say, a large corporation with many offices, so that the PABXs work concertedly to present the illusion of one large PABX. For example, given sufficient signalling facilities, call diversion from an extension on a PABX in London to an extension on a PABX in Glasgow could be possible. The *digital private network signalling system* (DPNSS) addressed this possibility. Based on DASS 2, it supported both the usual call control signals and some specialized ones for advanced call management. DPNSS was designed to use time slot 16 of an E-1 trunk, and like DASS 2 was incompatible with the physical layer protocol I.420 which now prevails. DPNSS is therefore being superseded by the ETSI protocol *unified international digital corporate network signalling standard*, or Q.Sig for short.

V5

The wide range of access network protocols (DTMF, loop-disconnect, ISDN and more) presents a problem for switch designers. To address this problem, *access network* (AN) equipment is often designed to convert the user's signalling system to a standard *local exchange* (LE) signalling system, at the AN-LE interface (the V reference point), as shown in Figure 6.3. The standard signalling system is called V5.

V5 defines the following protocol elements:

- a frame structure (LAP-V5–EF);
- a data link protocol (LAP-V5–DL);
- a PSTN application layer protocol (V5–PSTN) to communicate PSTN signalling events;
- a control protocol (V5–control).

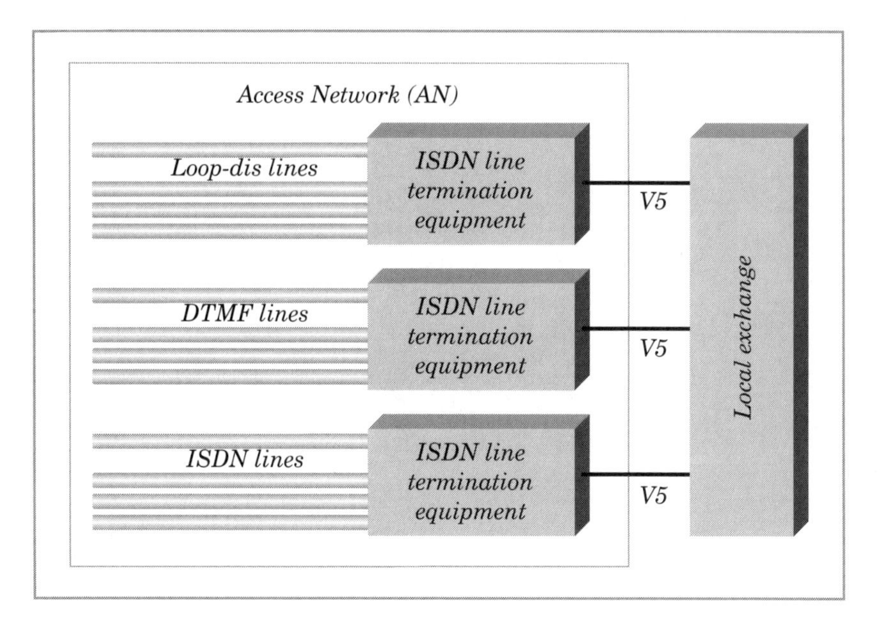

Figure 6.3 V5

V5 exists in two versions, V5.1 and V5.2. V5.2 adds several features:

- support for ISDN primary rate access;
- a bearer channel control protocol (V5–BCC) to dynamically allocate and deallocate bearer channels to calls, under the LE's control;
- a link control protocol (V5–link–control), for multilink AN–LE signalling interfaces;
- a protection protocol (V5–protection).

There are also standards for a 'VB5.1' reference point, between a digital service node and a broadband (or combined broadband and narrowband) access network.

V5 standards

Both ETSI and the ITU-T define V5 standards (identically), as follows:

Scope	ETSI	ITU-T
V5.1 interface standards	ETS 300 324	G.964
V5.2 interface standards	ETS 300 347	G.965
VB5.1 interface standards	EN 301 005 & 217	
Management of V5 interfaces	ETS 300 377–379	

SS7

Signalling between circuit switches within telco networks (and between networks) is now dominated by *signalling system number* 7 (SS7 or C7). SS7 is defined in a series of ITU-T recommendations (Q.700 series), and also in a set of ANSI standards (T1.111 *et seq.*), which are technically similar, but different enough to be incompatible.[3] All large-scale circuit switches support SS7 interfaces, and the system is very likely to become universal within its sphere.

SS7 offers many improvements over previous trunk signalling systems such as 10 pps, MF and SS6:

- its messaging offers high information content, to support value-added services;
- it is functionally rich, offering far more than simple call control;
- it has specific features to support IN (discussed further in Chapter 7);
- it supports redundant signalling network architectures, which offer high service availability;
- it is fast, giving low end-to-end delay.

In fact, it was this last feature that had the most obvious effect for users when SS7 was introduced in the late 1980s. Before SS7, a

[3] The differences principally concern the format of routing labels in SCCP.

trunk call would involve a delay, often of several seconds, while the address information was slowly propagated across the network. Then one day, users found their calls going through more or less immediately when they entered the last address digit. That was the introduction of SS7.

Before SS7, making a trunk call would involve a delay, often of several seconds.

Protocol model

Figure 6.4 outlines the main elements of the SS7 protocol suite. The definition of SS7 began before the OSI seven-layer reference model was standardized, and it was done at a time when the convergence of telecommunications and computing was less well established. Therefore SS7 does not follow the OSI model (or any other general model, for that matter). There are no explicit transport, service or presentation layers; the most that can be said is that the SCCP (described below) provides a service which more or less corresponds to the OSI network service.

The four lower layers of the stack, MTP levels 1–3 plus SCCP, are concerned with getting messages across an SS7 network

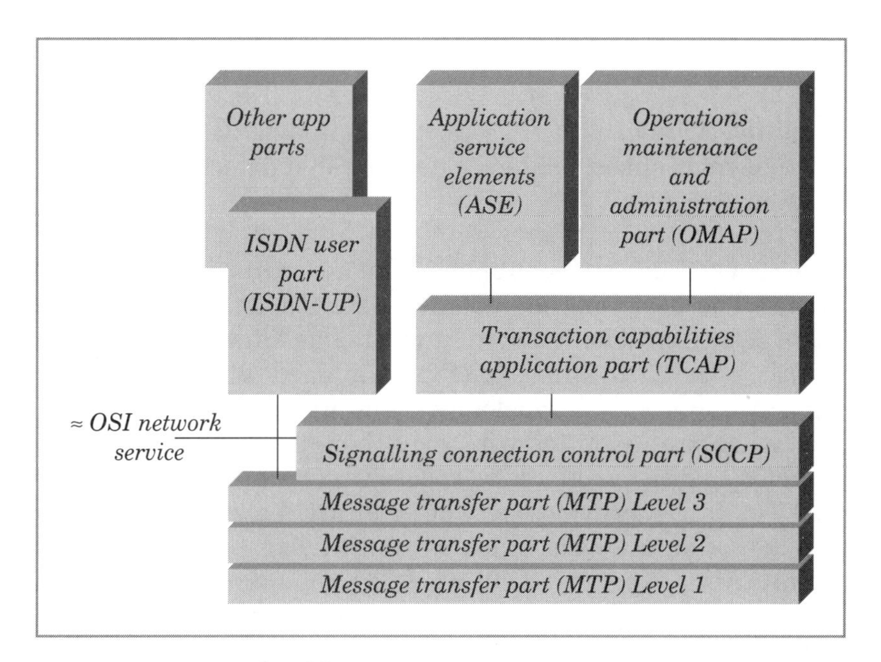

Figure 6.4 SS7 protocol model

(which may contain large numbers of nodes, and be a substantial data communications network in its own right). The upper layers (called application parts) define message sets for a range of application areas.

Message transfer part (MTP)

The MTP is defined in ITU-T recommendations Q.701–Q.710. MTP level 1 defines the physical layer protocol, which mandates a bi-directional data communications facility, realized as:

- 64 kbps on a E-1 bearer (ITU-T); or
- 56 kbps on an T-1 bearer (ANSI); or
- a modem link.

MTP level 2 defines an HDLC-style data-link protocol, including link-level error detection and retransmission.

The MTP level 2 frame types are:

- *message signal unit* (MSU – carries up to 272 octets payload);
- *link status signal unit* (LSSU – contains no payload, just link status flags);
- *fill-in signal unit* (FISU – keeps the link active, to maintain link synchronization and to enable rapid detection of link failure).

Error management is supported by two alternative mechanisms. The basic method operates like standard HDLC, with MSU retransmission only on request. The alternative, *preventative cycle retransmission* (PCR), is used when the link is subject to long round-trip delays (as on a satellite link). In PCR, MSUs are retransmitted whenever there is a gap in the traffic, just in case their originals were lost or corrupted.

MTP level 3 provides network-routing facilities. Within an SS7 network, a device that has routing capability is called a *signal transfer point* (STP). An STP may be part of a circuit switch, or it may be a stand-alone box. An STP will route messages across data links, including a facility for managing load sharing between multiple parallel data links between the same two STPs. STP routing is based on the message's *destination point code* (DPC) and originating point code (OPC), which are network level addresses, and on a *signalling link selection* (SLS) field.

Signalling connection control part (SCCP)

The SCCP, defined in ITU-T recommendations Q.711–Q.716, builds on the MTP's facilities. It extends the addressing, translating between globally valid addresses (such as E.164 telephone numbers) to locally significant network addresses. It also offers a range of classes of service:

- 0: basic connectionless
- 1: sequenced connectionless
- 2: basic connection-oriented
- 3: flow control and connection-oriented.

SS7 network architectures

SS7 signalling networks are of great importance to the telcos that operate them, and the science of SS7 network design is well developed. Generally, the topology of an SS7 network will not be the same as the bearer network that it supervises. SS7 network architectures may include elements of highly connected mesh, or high-capacity backbone with tributaries, to achieve suitable levels of throughput, resilience, and of residual capacity in the event of failure.

Generally, the topology of an SS7 network will not be the same as the bearer network that it supervises.

The standards bodies set performance standards for SS7.[4] Example criteria are:

- Availability: MTBF 1.9×10^5 s per signalling route
- Dependability:
  - Undetected errors < 1 in 10^{10}
  - Lost messages < 1 in 10^7

[4] Another point at which the ITU-T and ANSI standards diverge.

  - Data link bit error rate < 1 in 10^6
  - Message sequence error rate < 1 in 10^{10}
- Delay – of the order of 40–100 ms per STP

From the software engineer's point of view, SS7 work is demanding firstly because of the very high throughput required, and secondly because of the very severe system integrity requirements, with availability typically required to exceed 99.999 per cent.

ISDN user part

The *ISDN user part* (ISDN-UP, often collapsed to ISUP), defined in ITU-T recommendations Q.761Q.768, is one of several user parts that have been defined.[5] It supports both voice and (limited) data call facilities, and is now used in most telephony networks, regardless of whether the users have ISDN connections. It is specifically designed to interwork with Q.931 D-channel protocol discussed above.

As well as a basic bearer service, the ISDN-UP supports supplementary services including:

- user-to-user signalling (astonishingly, in three flavours);
- *closed user groups* (CUGs);
- *calling line identification* (CLI), otherwise known as *automatic number identification* (ANI);
- call forwarding (all calls, or on busy, or on no reply).

Transaction capabilities application part (TCAP)

The TCAP, defined in ITU-T recommendations Q.771–Q.775, provides a toolset to support remote application invocation across the SS7 network. TCAP does not specify what these applications (called *application services elements*) are. Typical uses of TCAP include remote access to route databases, and the co-ordination of IN services across multiple switches.

[5] Others include the obsolescent *telephony user part* (TUP) and *data user part* (DUP), and the *mobile application part* (MAP) described in Chapter 4.

ISDN-UP message types

Set-up

IAM: initial address message
SAM: subsequent address message
FAM: final address message
IFAM: initial and final address message
ACI: additional call information
SIM: service information message
SNM: send N digits
SASUI: send additional set-up information
ASUI: additional set-up information
ACM: address complete message

Answering

ANS: answer
RE-ANS: re-answer

Failure

CON: congestion
TERM CON: terminal congestion
CNA: connection not admitted
SUB-ENG: subscriber engaged
SUB-OOO: subscriber out of order
SUB-TFRD: subscriber transferred
RA: repeat attempt

Teardown/Release

REL: release
CLEAR: clear
CCT-FREE: circuit free

Information transfer

UUD: end-user data
NEED: nodal end-to-end data

Miscellaneous

SWAP: swap voice/data
BLKG: blocking (take channels out of service)
CFC: coin and fee checking
HLR: howler
TKO: trunk offer

Structurally, TCAP comprises two sub-layers:

- a *transaction sub-layer*, which delimits dialogues, rather like the OSI session layer;
- a *component sub-layer*, which does message bundling, sequencing and request/response structuring. This sub-layer is very like the OSI *remote operations service element* (ROSE), extended to accommodate large responses.

Operations maintenance and administration part (OMAP)

OMAP addresses the issues of managing an SS7 network. It defines protocols and procedures to monitor, co-ordinate and control the network, including standard management functions such as the *MTP routing verification test* (MRVT).

I must confess, however, that although I have read a good deal about OMAP, he has never actually worked on a serious implementation of it. There are other, more general, network management technologies (see Chapter 11) which may be preferable.

Further reading

Many of these signalling systems are covered well by Flood (1997) but Hiett (1991) is a slightly more reader-friendly alternative, if you can get hold of it. Cole (1999) adds detail from a US perspective. SS7 is presented very well in Modarressi and Skoog (1990).

7 Circuit mode switching

Sepulchral introduction

In the early days of voice telephony, the network operator's staff did switching between customers manually. The first automatic circuit switching equipment was developed by a Mr Almon Strowger. Mr Strowger, so the story goes, was by profession not a telecoms engineer but an undertaker in Kansas City in the US, in the late 1800s. In this town there were two undertakers, and Mr Strowger's rival had a key advantage: he was in league with the local switchboard operator. When anybody died, the bereaved would call one undertaker or the other, but no matter which one was asked for, the operator would put the caller through to Mr Strowger's rival. This, allegedly, was the spur that led to the development of the first automatic telephony switch, which Mr Strowger patented in 1889.

Be that as it may, the invention was not implemented on a large scale until the 1920s, when it became part of the telcos' attempts to thwart the labour power of their workforce of manual operators.

Mr Strowger's switches were electromechanical, and closely integrated with the 10-pulse-per-second signalling system. Each pulse train would drive relays and actuators to move wiper blades across arrays of electrical contacts laid out in rows (called *levels*) and columns. Typically, each array would handle two successive digits. By cascading such arrays, the incoming digits could be made to set up the call path across the switch.

Strowger exchanges were slow, large, unreliable and power-hungry.

Strowger exchanges were slow, large, unreliable and power-hungry. They also necessitated a fixed relationship between telephone numbers and physical connections, which led to strange and inconsistent numbering plans. A significant improvement on the basic Strowger design was the director system (alias register addressing), which used a system of digit registers to store incoming digits and then translate the digit string (with a look-up table) before acting on it. The director system therefore allowed some improvements in the consistency of numbering plans.

Strowger technology was partly superseded by another electromechanical system called crossbar. Patented in 1918, crossbar switches separated the call-control equipment from the switch fabric, and so could get by with less call control equipment than a Strowger switch. Electromechanical systems were in turn partly superseded by a hybrid electronic/electromechanical system called reed relay, where an electronic call-control system operated small relays which made the circuit connection.

However, all of these systems were rapidly displaced in the 1970s by the arrival of stored program control electronic switching systems, as described here.

Now that general-purpose computers are very small, it can be a surprise for a software engineer to see a large circuit switch. When the first computers were developed, their 'main frames' were derived from telephony switch hardware. Because of the need for huge numbers of interface cards to connect to typically low-rate user circuits, telephony switches are still very impressively big things, reminiscent of launderettes (but more costly).

A note on nomenclature: *circuit mode switches* are sometimes called just that, but they are also called *switching nodes* (by network planning people), *central offices* (CO) by Americans, and *telephone exchanges* (by most of the rest).

Electronic switching systems

To understand how a circuit switch works, it helps to approach it as two parts:

- the switching fabric (or switch matrix), the part that gets the traffic from one user port to another;
- the call control (the part that manages what connections the switching fabric should make).

Switch fabrics

Circuits typically arrive at switches as E-1/T-1 digital circuits. Sometimes higher-rate interfaces are used, sometimes alternative digital interfaces such as ISDN, and sometimes analogue connections (to codecs built into the switch's interface cards). But E-1/T-1 connections predominate, and also serve to illustrate the issues of digital time domain switching, which is the core technology for all modern circuit switches.

Advantages of digital time domain switching

Digital time domain switching means switching in time (i.e. from one time slot of a TDM circuit to another) as well as in space (from circuit A to circuit B). It is attractive for three main reasons. Firstly, digital switching can be done with much simpler transistor circuits than analogue requires. Secondly, switching between TDM multiplex circuits removes the need to demultiplex down to single circuits to connect to the switch. This saves on demultiplexers, and allows the switch to be smaller and consume less real estate. Thirdly, it makes advanced call handling easy. For example, a one-way circuit can be implemented by only copying the data in one of the directions. Or again, conferencing (which used to require difficult analogue circuits to mix the signals) can be implemented through arithmetic logic circuits.

Time and space switching

A digital time domain switch fabric has to connect the time slots of the incoming circuits to the correct time slots of the correct outgoing circuits. The architectures for doing this are many and complex. They all, necessarily, work by combining elements of time switching and space switching. Time switching means reading the data out of one time slot on a TDM channel, and writing it into another one on the same channel. This requires storage registers and close control of delays. Space switching means taking data from one TDM channel and sending it to one of many other TDM channels, and usually also changing channels very rapidly.

Space time space architectures

In the early days of digital time domain switching, time switches were new and costly, whereas space switches were relatively cheap. Therefore early switches used a *space time space* (STS) architecture, as illustrated in Figure 7.1. This figure shows a very simple STS switch, which switches between N (N ≤ 30) incoming circuits and N outgoing circuits, for each of 30 time slots per circuit. In real life, the call paths will generally be bi-directional, and so the outgoing and incoming circuits will not generally be distinguishable, but it helps to separate them for showing how the switching works.

The figure shows a switch connecting time slot 3 of incoming circuit number 2, to time slot 6 of outgoing circuit number 1. To achieve the time shift, a 3-slot delay unit is used. A space switch at the incoming side switches circuit 2 through to the delay unit, just for the duration of time slot 3. A space switch at the outgoing side switches the output of that delay unit through to outgoing circuit 1, for the duration of time slot 6. The space switches are controlled by blocks of memory called *connection stores*, which are written to by the call control software.

However, the economics of switch design are different now.

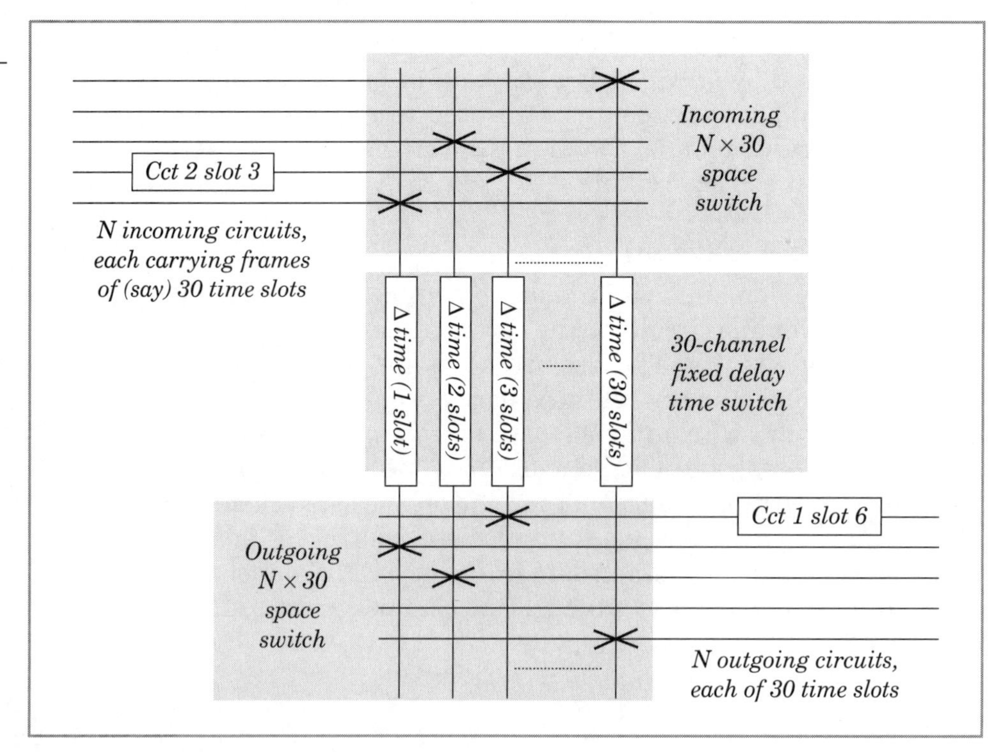

Figure 7.1 STS architecture

Time switching is readily (therefore cheaply) achievable in integrated circuits. Space switching, not least because it requires physical interconnect to many channels, is more costly to implement than time switching. Therefore, modern circuit switches generally use some variant or elaboration of the *time space time* (TST) architecture outlined below.

Time switching is readily (therefore cheaply) achievable in integrated circuits.

Time space time switching

Figure 7.2 shows a simple TST architecture, addressing the same switching problem as Figure 7.1. At its heart is an N × N space switch, which cycles all the possible N^2 combinations of space connections once per frame time. The incoming circuit (2) is permanently connected to a variable delay element, which is programmed to switch slot 3 into the time slot for which the space switch connects incoming circuit 2 to outgoing circuit 1. Then a second delay element switches that time slot on circuit 1 into slot 6, as required.

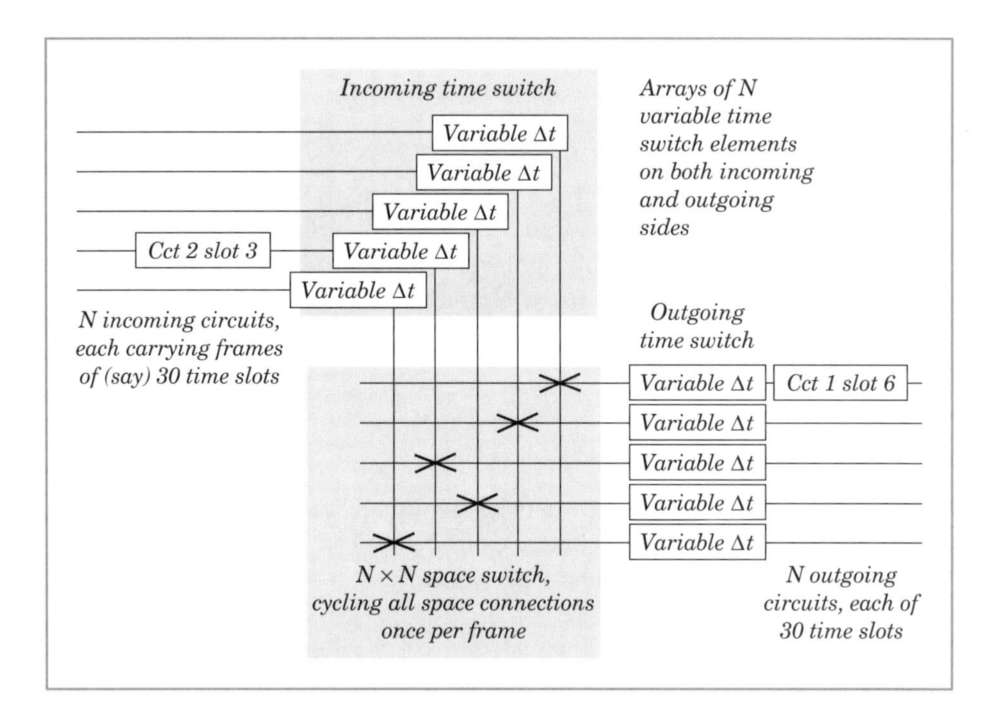

Figure 7.2 TST architecture

Blocking

In both of the above examples, there were possibilities of blocking; that is, of failing to have enough switching resources to make all the required connections. In Figure 7.1, the 3-slot delay unit might have already been occupied by some other call in the required time slot (number 3). In Figure 7.2, some other time slot from incoming circuit 2 might already have been using the one available connection time for connecting to outgoing circuit 1.

In practice, a switch architecture with the potential to block calls may be acceptable, because the distribution of traffic over time (and over the circuits) may make the probability of blocking acceptably small. However, in some cases (for example, for emergency services call centre switches), blocking has to be eliminated, and in all cases it has to be minimized. In an STS system, blocking can be minimized by providing additional sets of delay units. In a TST system, the same effect can be achieved by increasing the speed of the internal TDM highways passing through the space switch, so that it cycles through the possible connections several times per frame time. In this case, the variable delay units also have to adapt the high-speed internal highways to the external circuit rates.

More complex architectures

The very simple architectures shown in Figures 7.1 and 7.2 are not suitable for scaling up to the thousands of circuits handled by real public network exchanges. In practice, more complex combinations of time and space switching elements are used, to achieve a balance between capacity, blocking and economy.

Uses of switch fabric without call control

Circuit switch fabrics can be deployed without associated call control systems, in the cases of *automatic cross-connect equipment* (ACE) and *concentrators*, as described below.

Automatic cross-connect equipment

An ACE (also often called a *digital automatic cross-connect system* or DACCS) is used where there is a requirement for very slow switching. That is to say, where a circuit has a lifetime of days or years, and where it can be set up and cleared down at leisure. An ACE unit has a number of incoming and outgoing TDM circuits, and can be configured (manually via a craft terminal, or via a network management system) to route particular time slots from incoming channels to particular time slots on outgoing channels. ACE exists in both PDH and SDH/Sonet forms.

ACE is typically used for:

- *switching* leased lines between subnetworks, as shown in Figure 7.3;
- *consolidation* at the interface between the access network and a circuit switch, as shown in Figure 7.4. Where an access network uses TDM transmission, patchy service uptake by subscribers may mean that many of the TDM circuits that arrive at the switch are only partly used; a significant part of their capacity may not be allocated to any customer's network connection. Circuit switches are costly, and a significant cost driver is their number of TDM ports. ACE can be used to consolidate the connections, so that a large number of poorly filled TDM ports is converted to a smaller number of well-filled ones. The saving in switch cost more than justifies the cost of the ACE;

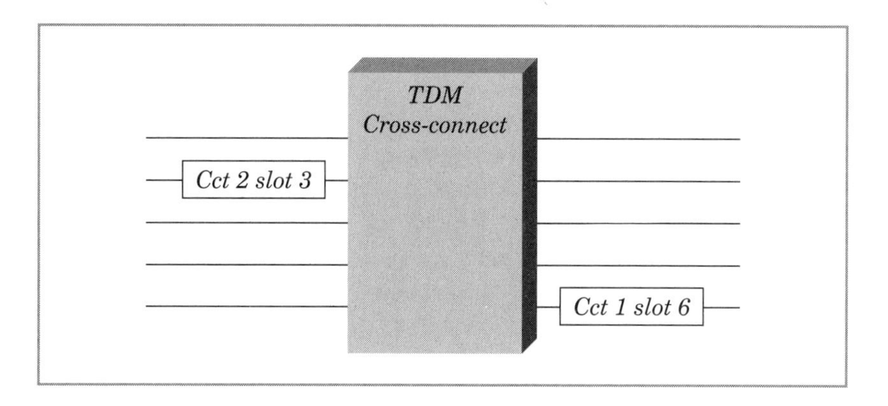

Figure 7.3 Switching leased lines

- *grooming*, which is similar to consolidation. Grooming provides a consolidation function, but also an element of routing, so that a number of destination systems can be served. For example, a grooming system may split out PSTN circuits from IP data circuits;
- *interconnect* between TDM subnetworks, as shown in Figure 7.5 and discussed in Chapter 5.

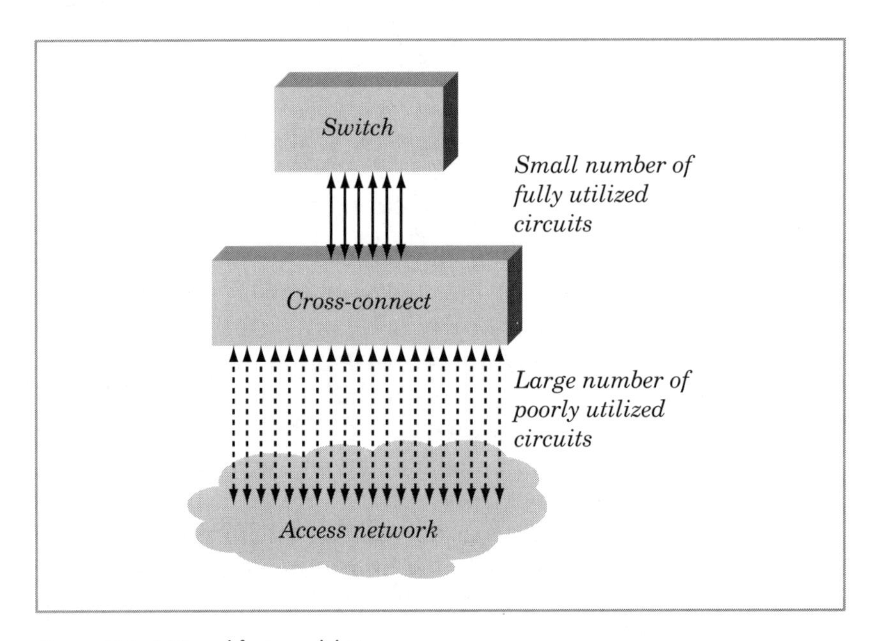

Figure 7.4 ACE used for consolidation

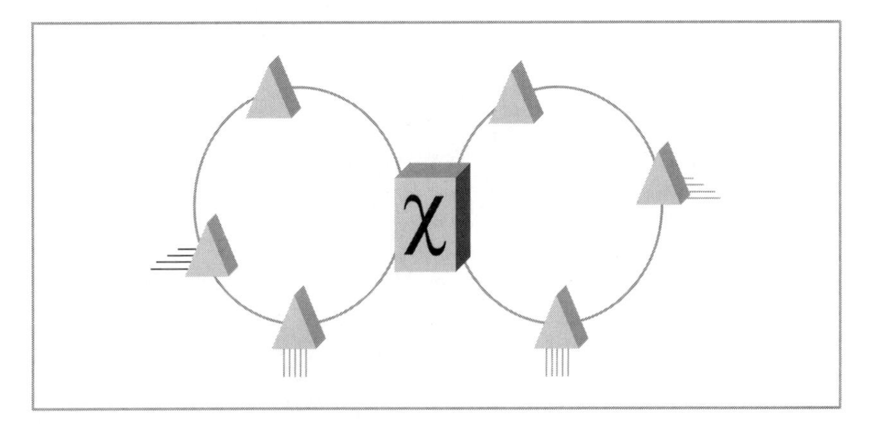

Figure 7.5 Connecting subnetworks

RCU/Concentrators

Concentrators are like ACE, but more subtle. We have seen how an ACE can be used as a consolidator, to weed out transmission capacity that is not actually connected to a customer terminal. But even then, of the customer terminals connected to a switch, only a few will be active at a given time. Concentrators, also called *remote concentrator units* (RCUs) or *dependent units*, operate as satellites to a circuit switch, and just connect active terminals, not the *n* per cent idle lines.

The use and value of concentrators can be seen through consideration of a typical wireline local network. Such networks, in many countries, were typically set up when circuit switching was electromechanical. The cost of the local loop, and large size of these switches per line served, meant that in a large town there would usually be one large switch in the centre, and a number of satellite switches in the suburbs. The advent of modern digital switching meant that the town could be served instead by a single powerful switch in the centre, and this would bring economies of scale as well as operational efficiencies.

The advent of modern digital switching meant that the town could be served by a single powerful switch in the centre.

However, the local loop would still be terminated in the old buildings that housed the suburbs' switches. To extend the local loops all the way to the town centre would be outrageously costly and often impracticable. To install codecs on all the local loop terminations and bring all the digitized signals back to the centre would also be wasteful. The solution is to install concentrators at the ends of the local loops. The concentrators manage the provision of dial tone, the allocation of codecs and of digit detectors, and pass through only the active callers.

Basic call control

The basic functions of call control software are simple, and relate very closely to the basic signalling events presented in Figure 6.1.

Seize detection looks for the call request/seize signal. This triggers the allocation of digit detectors to the incoming circuit and generation of dial tone (a switch is not generally equipped with enough digit detectors or dial tone generators to serve all its users simultaneously). A *digit analysis* function acquires the called address, which is then used by a *routing* function to select the correct outgoing circuit. Routing is generally done by more or less complex look-ups in routing tables. Routing arrangements are discussed further below. Assuming for simplicity that the outgoing circuit is on the same switch (as in the case of a local call), the switch then generates the alerting signal and allocates seize detectors to identify the moment of answering. A *call path establishment* function then (and only then) invokes the switching fabric to establish the bearer connection.

Other switch features

Circuit switches offer many features beyond basic call control. Some of the more significant ones are:

- *nailed-up connections:* many switches can be configured to support semi-permanent circuits, just like the ACE described above. However, when done through a circuit switch, these are generally called nailed-up connections, rather than leased lines;

- *CLASS/LASS features:* some US telcos have recently defined a standard collection of advanced user facilities, under the label CLASS,[1] or alternatively LASS.[2] While the features in the list are individually unremarkable (they have all been available for some time), the list is significant in being the target which many telcos are now aiming for.

CLIP and CLIR

Switches in SS7 signalling networks have the possibility of receiving from the network, and presenting to the called party, the address of the calling party. This is *calling line identity presentation* (CLIP,[3] standardized in ITU-T recommendation I.251.4). *Calling line identity restriction* (CLIR,[4] standardized in ITU-T recommendation I.251.4) is the facility where a network allows the calling party to withhold his or her address.

[1] Custom local area signalling Services (AT&T term).

[2] Local area signalling services (Pacific Bell term).

Call blocking

Call blocking (or *selective call rejection* – SCR) allows each customer to control a list of telephone numbers from which they do not wish to receive calls. Any calls to one of a customer's specified numbers are blocked within the network.

Call return[5]

When a customer (A) calls another party (B) who is engaged, A can ask the network to call A back when B has become free. Then, when A picks up his or her phone, the network automatically calls B again (if B is still free).

Call trace

Call trace allows a called party to retrieve the time and originating number of the last incoming call.

Priority ringing

Priority ringing (or *distinctive alerting*) offers a distinctive ringing cadence when the calling number is in a user-controlled list.

Repeat dialling and call return

Repeat dialling (confusingly also called *automatic recall* or *automatic callback*) lets the customer call (with a single keystroke) the last number that they dialled. Call return lets them call the last number that called them.

Selective call forwarding

Selective call forwarding (SCF) automatically forwards calls to a chosen destination, but only if the calling number is in a user-controlled list.

[3] CLIP is also called individual calling line ID (ICLID), Number ID and Calling Number Delivery.

[4] CLIR is also called Privacy, Number ID Blocking, and Calling Number Delivery Block.

[5] Alias callback or recall.

Hunt groups

A switch may provide a hunt group routing facility where calls addressed to a single address are routed to any of a set of addresses, typically on a round-robin basis.

Centrex

Increasingly, public networks can offer advanced call processing features such as hunt groups, call waiting indication and ring back when free, which were previously the province of PABXs (see below). Such facilities, when offered by a public network, are called Centrex. They are sometimes implemented in the network's circuit switches, but often it is more economical to implement them through cheap front-end switches (between the main switches and the customers), or via *intelligent networks*, as also described below.

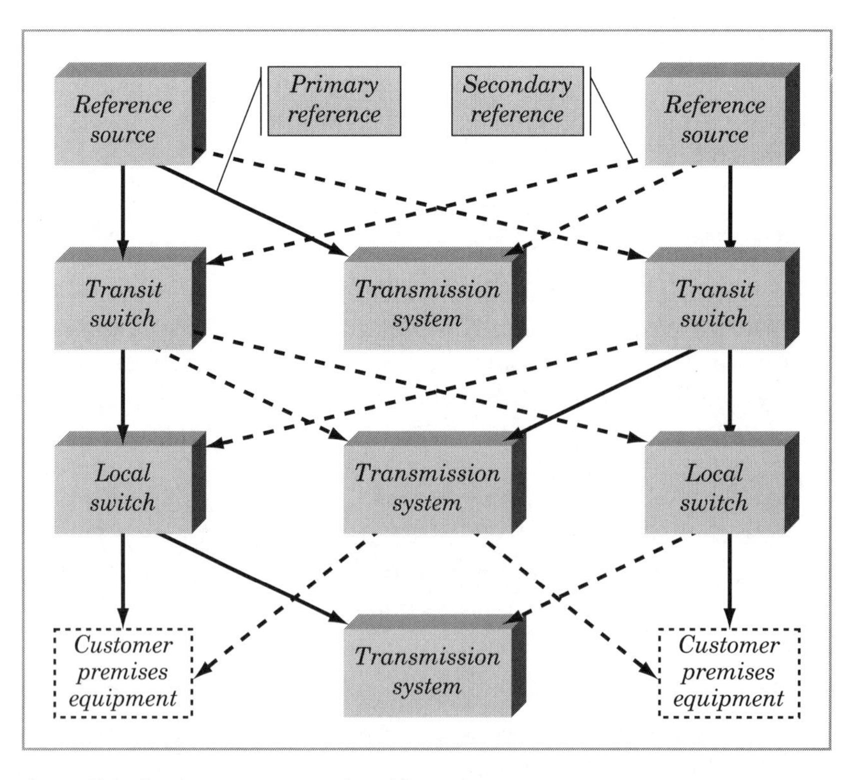

Figure 7.6 Synchronization network and hierarchy

Switches and circuit mode network synchronization

Many switching functions, and particularly the time switching of digital circuits, require precise synchronization between the switches of a network. There is therefore, in any operator's network, a switch synchronization overlay network. Typically, as shown in Figure 7.6, there will be dual redundant primary references (atomic clocks), serving a hierarchy of exchanges and transmission systems, arranged so that the timing drift across the network is minimized, to the order of 1 part in 10^{11}.

Current large-scale digital switch products

Generally, this book avoids naming specific products. However, the names of the major circuit switch products are so ingrained into telecoms culture that the knowledge of them is a necessary survival tool. The following list therefore is not an attempt to list every current digital switch product, but to introduce the ones that have become part of the common culture.

- 5ESS, originally from AT&T, now offered by Lucent Technologies.
- DMS 100 (and 200, 250 etc.), from Nortel (formerly Northern Telecom and Bell Northern).
- EWSD (which stands for Elektronisches WählSystem [Digital]) from Siemens.
- System X, one of the earliest SPC switches, originally built for BT by GEC and Plessey, and now owned by Marconi.
- AXE-4 and AXE-10, from Ericsson (AXE-10 is also called System Y in the UK, as if it were closely related to System X, which it isn't, although its architecture is in fact quite similar).
- E10 & E12 from Alcatel.
- NEAX from NTT.

While new switch products keep on emerging, the immense software investment in the systems listed above means that they are unlikely to disappear in a hurry.

While new ones keep on emerging, the immense software investment in the established switch products means that they are unlikely to disappear in a hurry.

PABXs

In parallel with the evolution of public network SPC switches, smaller switches have been developed for use within a single organization, and usually within one business location, rather than offering service to the public. These are called *private automatic branch exchanges* (PABXs), sometimes, more loosely, *private branch exchanges* (PBXs), and even occasionally *computerized branch exchanges* (CBXs).

A PABX will support between a few tens and a few thousands of telephones, and can thus be larger, in some cases, than a small public network switch. A PABX is connected to one or more public networks, via either multiple analogue voice circuits, or primary rate ISDN, or E-1/T-1 circuits controlled by SS7 signalling. Typically a PABX engineer will call such circuits *trunks* (whereas a public switch engineer would expect a trunk to be a substantial aggregation of circuits).

PABX features

While the switching fabric of a PABX is usually a simplified and scaled-down version of that in a public network switch, a PABX will often offer comparatively advanced call-control facilities. This technical lead has been encouraged by the nature of the PABXs' operating environments:

- the very short range of the access network has allowed deployment of proprietary multi-pair interfaces, supporting terminals ('featurephones') with a rich user interface capable of invoking advanced services;
- the small system scale has permitted feature designs that would be unworkable in a large network. For example, complex call-divert patterns, which are immensely difficult to organize across a large public network, can be

implemented subject only to the constraints of the PABX's programming environment.

Table 7.1 introduces a range (not by any means exhaustive) of the more common PABX features. Here 'extension' is used as the usual name for a PABX user terminal.

Feature	*Effect*	
Making calls	*Abbreviated dialling,* also called repertory dialling and speed dial	A short digit sequence keyed by the user on his or her extension, or the pressing of a single key, is translated by the PABX into the full outgoing number. The translation may be either specific to the user's terminal, or common to all users of the PABX.
	Camp-on	A user on a current call is made aware that a third user wishes to call him or her.
	Intrude (executive override)	Interrupts a current call to give a specially privileged user immediate access to one or both of the callers. This may either be overt (for example, where a manager can interrupt his subordinates' calls) or covert (for staff monitoring). Use of covert intrusion facilities raises difficult ethical and legal issues, and must not be done lightly.
	Call back when next used	User A calls user B, who doesn't answer. User A invokes the callback feature via a short-digit sequence, then replaces the handset. When B next uses the

Table 7.1 cont.

Feature		Effect
		telephone, at the end of the call the PABX calls A (usually with a distinctive ringing cadence). If A picks up the telephone, the PABX then calls B (unless of course B has meanwhile begun another call).
	Call back when free	User A calls user B, who is using the telephone already. The PABX presents 'engaged' tone to A. A then invokes callback (as above), and replaces the handset. When B's call completes, the callback process described above is triggered.
	Conference	Three or more parties (who may be a mixture of directly connected PABX users and external callers) are connected in a single conversation.
	Direct outward dial (DOD)	A user can make calls across the public network (typically by pressing the '9' button) without having to ask an operator.
	Intercom (alias 'speak')	A user can speak directly to another user (via a loudspeaker feature built into the PABX telephones) without the receiving user taking any action to answer the call.
Receiving calls	*Direct dial in (DDI) or direct inbound dialling (DID)*	By arrangement with the relevant public network operator, the PABX's users are allocated a block of telephone numbers which

Table 7.1 cont.

Feature		Effect
		share a common prefix: for example, +44 1225 475xxx. The local public exchange routes calls to all numbers in that block to the PABX, and passes through the remaining digits. These digits are used by the PABX to route the calls to the correct extensions.
	Automated attendant	Incoming calls are passed to an IVR (interactive voice response) system, which collects information from the callers (via speech recognition or MF tones) and routes it to suitable destinations.
	Call forward or divert	Calls are diverted to the nominated extension number, either unconditionally or on no answer or on busy. Implementing a divert system can be a significant software challenge; firstly because of the complexity of a multi-user divert system, where for example potentially circular divert patterns can be set up; secondly because of the interaction of the call divert with other call-control features, for example callback.
	Follow me	A user (say 'A') can divert their calls to user B's extension. Then later, from B's extension, A can re-divert his or her calls to C's extension, and so on.

Table 7.1 cont.

Feature	*Effect*	
	Distinctive ringing cadence	Where several users share an office and a telephone instrument, they can be assigned separate extension numbers, and each number will cause the telephone to ring with a different pitch or rhythm.
	Night service	At the end of the working day (either manually or automatically), a secondary call distribution map is applied. This may for example route incoming calls to a night watchman or voice mail.
	Hunt group answering	Incoming calls for a chosen number are directed to any of a defined list of users. There are several heuristics used, including: • a simple system which always starts from the top of the list and works down until one is free; • uniform call distribution (UCD), which awards calls on a round-robin basis; • automatic call distribution (ACD), which awards calls so that each user takes a similar total duration of calls.
	Music on hold	As the caller holds, music plays either from aesthetically distressing internal tone generation equipment, or more satisfactorily from an external CD player.

Table 7.1 cont.

Feature	*Effect*	
	Call transfer	A user receives a call, then redirects the call to another extension.
	Pick-up groups	When a call arrives for a member of a work group, any member of the group can answer the call.
	Voice mail (vmail)	The PABX stores voice messages, usually on hard disk.
Hotel facilities	*Room maid*	As a hotel worker makes his or her round of the rooms, he or she enters a code at the extension in each room, so that the hotel management can verify that the complete round has been made.
	Room billing	When a guest checks out, the PABX works out, online, the call charges.
	Room enable	When a guest checks in, the extension in their room is enabled (when the room is not let, the extension is disabled to preclude use by hotel staff).
Management	*Call logging*	Most PABXs can record the time, duration and destination number of all outgoing calls. Some also allow users to allocate each outgoing call to a cost centre number, for subsequent charging to clients.
	Remote management access	A remotely sited management system can call a DDI number that connects to a modem within the PABX, and thence

	get access to the management facilities. An extreme case of this occurs where a PABX is leased, and the leasing company can remotely disable features if the bills are not paid.
Barring	Barring is generally available per extension. It can be applied to categories of call (external, national, international) as well as to specific PABX features (for example, *intrude*).

Table 7.1 PABX features

Key systems and hybrid systems

One alternative to a PABX is called a *key system*. In a key system, each of the trunks to the public network appears as a light and a key on each user terminal. Users can thus view the status of the trunks, and select a trunk for answering, or for an outgoing call (without having to dial '9'). As computer-based PABX technology has become cheaper, key systems have tended to be replaced with either ordinary PABXs or *hybrid systems*, which are PABXs with the addition of the direct trunk access features of a key system.

Automatic call distribution systems

Automatic call distribution is sometimes used simply to mean the capability of a PABX to route calls to its extensions without operator intervention. However, it is also used to describe the specific variant of PABX technology used in call centres, which is described further below.

Call centre technology

The key elements of a call centre (from a telecoms software point of view) are shown in Figure 7.7. The telephony interface of a call centre is built around an *automatic call distribution* system. An ACD is technologically like a PABX, but with a different balance between extensions and trunks, and a different call-control feature set, which is usually driven by programs and databases in an associated general-purpose computer. This integration of telephony switching and computing is called *computer telephony integration* (CTI).

Whereas a PABX will typically have far more extensions than trunks, and will have extensive features for calls between extensions, an ACD will have very limited (if any) inter-extension call facilities, and will often have more trunks than extensions. This difference in extension/trunk ratio is because a call centre focuses on serving as many external calls as possible, and on keeping its operators as busy as possible. Therefore incoming calls will be pre-processed, so far as possible, by voice response features, before connecting to an operator. Outgoing calls will be generated under software control (this is called predictive dialling), and will be connected to an operator only when the called party answers.

This difference in extension/trunk ratio is because a call centre focuses on serving as many external calls as possible.

The ACD will also have features not normally found on a PABX – for operators to enable and disable their position when

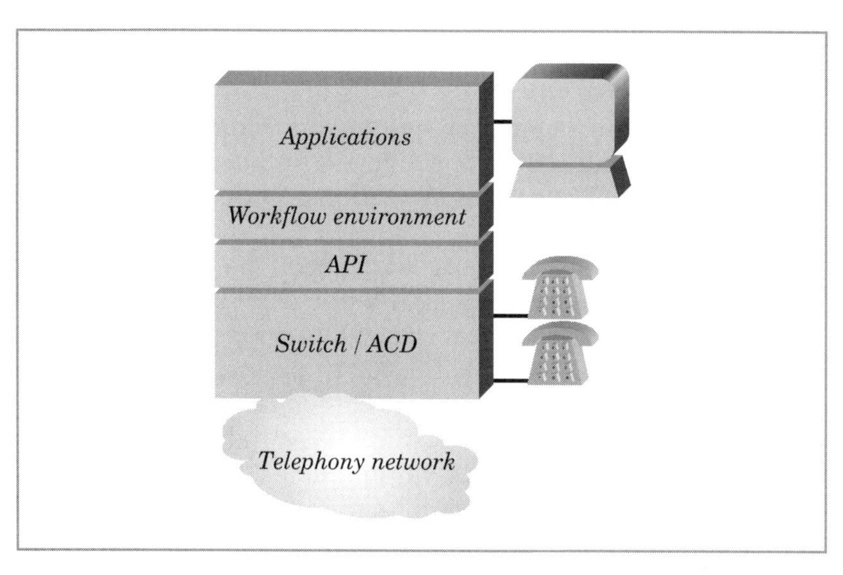

Figure 7.7 Call centre elements

they go on and off shift, and to temporarily block the distribution of new calls to them while they complete the work arising from the last call.

The interface between the ACD and the IT systems is through an API (*application programming interface*), and a number of functionally similar but technically incompatible APIs exist.

Call centre APIs

Setting aside a number of call centre APIs that are specific to ACD manufacturers, here are the best-known standard APIs for interfacing an ACD to a call centre application:

- TAPI (*telephony API*), defined by Intel and Microsoft, for Windows and NT;
- CAPI (*computer API*), defined by ETSI in ETS 300 838 as the API for ISDN call management;
- TSAPI (*telephony services API*), defined by Novell and AT&T, for NetWare;
- JTAPI (*Java telephony API*), defined by the Enterprise Computer Telephony Forum (which includes Lucent, IBM and others).

Call centre applications can be linked to incoming calls through both CLI and DDI. CLI is used to identify the caller and select the relevant data before the call is connected. DDI allows a single physical call centre to support a number of applications (often on behalf of several organizations), selected by the called number.

For both incoming and outgoing modes, the efficiency of the human operators is maximized by automatically presenting the relevant data for the call (called screen popping). For outgoing mode, the call centre application often presents the operators with a very closely specified script for the call. While this reduces the staff training required, it also reduces the interest of the work, and establishes a vicious circle. The combination of relentless workload and lack of autonomy can make the call centre operator's job both dull and stressful, and is one of the causes of the high staff turnover in call centres, that the scripting tries to address.

Telephony addressing

In a circuit-oriented telephony network, each network access point is identified by a unique address, i.e., its telephone number. Telephone numbers follow a hierarchical structure that is laid down in ITU-T recommendation E.164 (and 'E.164' is often used in speech as a synonym for 'telephone number'). E.164 (which is also published as I.331) imposes the following structure:

Field	Content
1	country code
2	area or destination code
3	local or customer number

E.164 aims to ensure that a normal international number, excluding the international access code, is not more than 12 digits long.[6]
Country codes
Country codes are based on *world numbering zones*, which are defined as follows:
1 North America
2 Africa
3 and 4 Europe
5 South America and Cuba
6 South Pacific
7 CIS (ex USSR)
8 North Pacific
9 Far & Middle East

Thus, for example, the country codes for Switzerland and Ecuador are 41 and 593 respectively.
Area or destination codes

Within each national administration, *destination codes* are defined independently. One example is the numbering scheme introduced by the UK regulator Oftel in 1999, presented in Table 7.2.

[6] It does however allow a further three digits' extension for ISDN subaddresses.

Codes	Use
0...	International access
1...	Geographic area codes
2...	Geographic area codes
3...	Reserved for expansion of geographic area codes
4...	Reserved for expansion of geographic area codes
5...	Reserved for corporate numbering (where an area code can be assigned to a corporation)
6...	Reserved for expansion
70...- 72..	Personal numbering (mobile and fixed)
76..	Paging
77..- 79..	Mobile
8...	Special services up to national rate tariff, including:
80..	Freephone
82..	Schools Internet
84..	Local rate
87..	National rate
9...	Premium-rate services, including
909.	'Adult' services

Table 7.2 UK numbering scheme

Network topology and routing heuristics

The shape of the network

Historically, telephony networks were strongly hierarchical, with very little connectivity between nodes at the same level of the hierarchy. For example, AT&T's analogue network in the US was built following the five-level hierarchy shown in Figure 7.8. This approach was driven by the large size of telephony switches, and the exorbitant cost of large-scale transmission trunks.

[7] The figure omits some concentrator units, which could perhaps be viewed as a third layer below the local exchanges, and also a derived services network, which certainly can be viewed as a layer above the transit exchanges. However, for purposes of comparison, these are irrelevant, as they are not considered in the example of Figure 7.8.

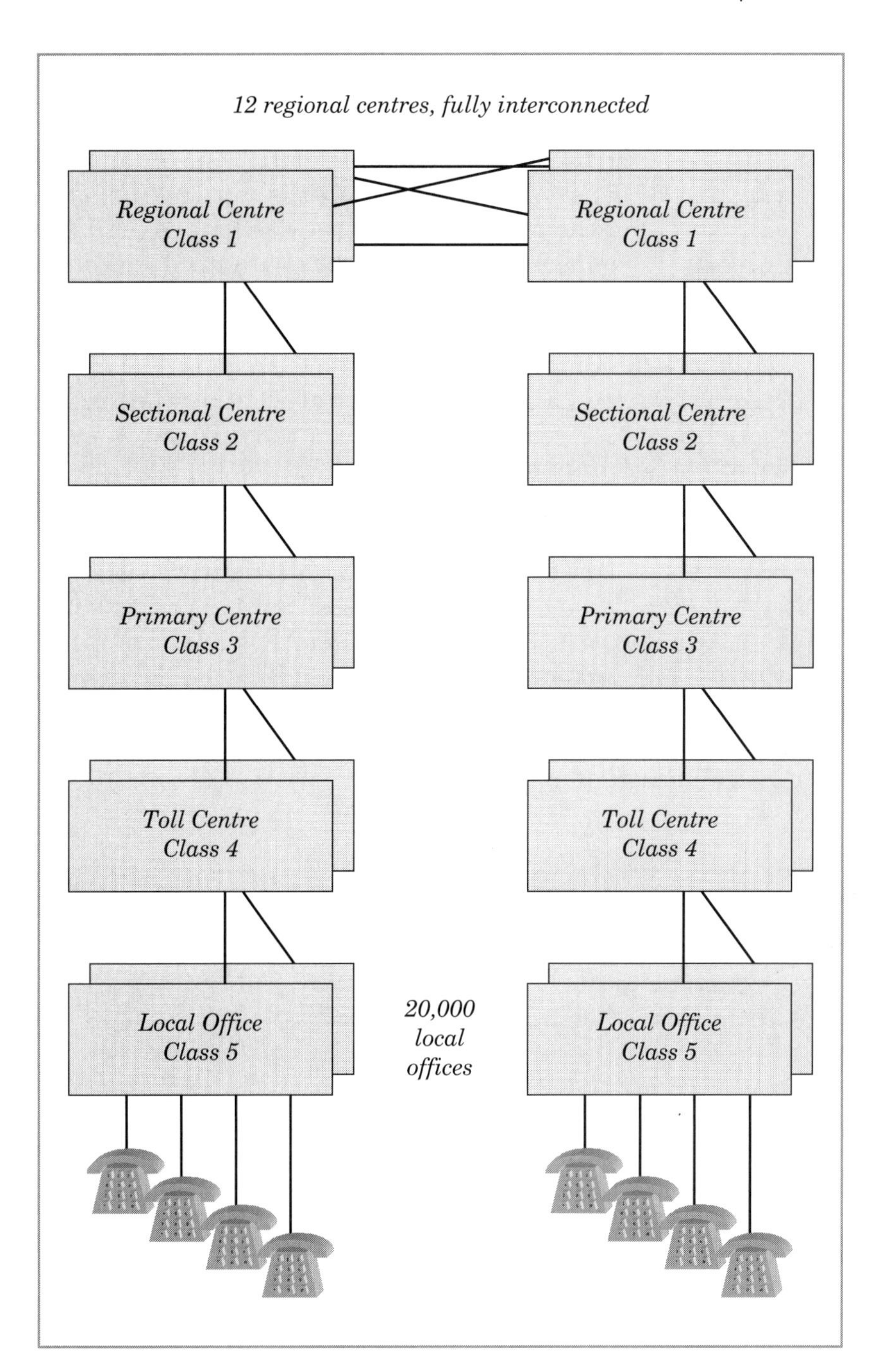

Figure 7.8 Hierarchical telephony network

The switches at the lower levels of the hierarchy were put there not least to concentrate traffic

The switches at the lower levels of the hierarchy were put there not least to concentrate traffic so that as little as possible would have to be back-hauled to the regional centres and the backbone network. Also, the routing capability of the analogue switches was limited, and full connectivity could not be guaranteed without a hierarchical plan.

As transmission costs have dropped radically since the 1960s, networks have become flatter and less hierarchical. For example, Figure 7.9 shows a simplified[7] view of BT's network in the UK, which has essentially a two-layer structure and is more typical of recent national network construction.

In this example there is a basic hierarchical network feeding traffic up to the *final routes* (also called *backbone routes*) between the tandem exchanges, and also a network of *direct routes* (or *transverse branches*) between exchanges at same level. As transmission costs are now very low, massive interconnect of trunk exchanges is possible, and BT's trunk network is fully interconnected. Also, all the local exchanges are *dual- or triple-homed*, i.e., connected to more than one trunk exchange (for load sharing and resilience).

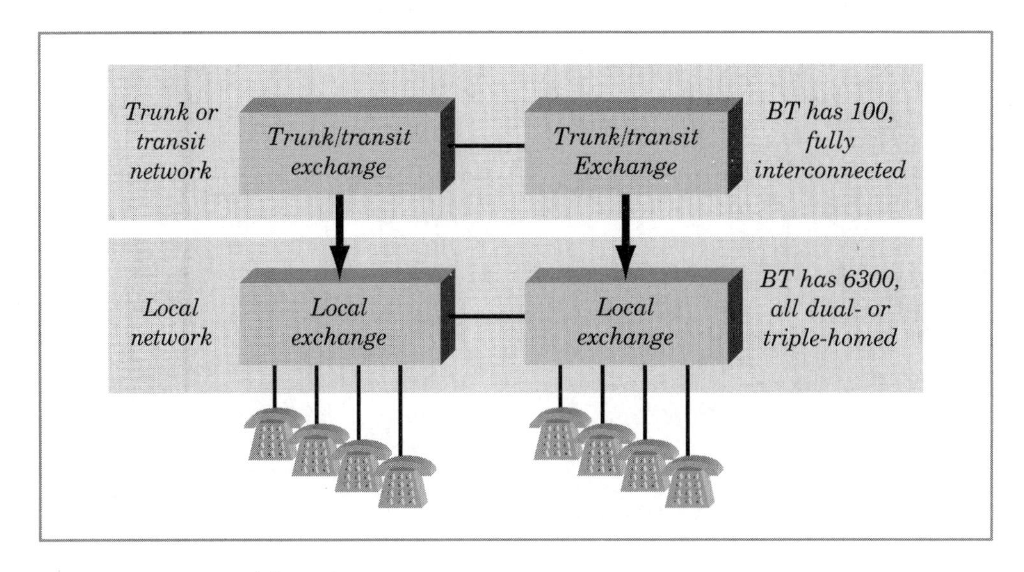

Figure 7.9 BT's network hierarchy

Routing at the exchange level

The routing function in a circuit switch is in principle simple. Each switch maintains a routing table, which maps directly connected subscribers' numbers to their switch ports, and which indicates, for all other numbers, which trunk the call should be directed to. In practice, routing is complex and the detail of routing tables is critical to a network's operation. The following sections present some of the more common subtleties of call routing.

Trombone working

A local area may be served by several small exchanges, all of which share the same area code. When a user in the area tries to call another user who is connected to the same exchange, the call request may be resolved directly by the local exchange. Alternatively, if the local exchange's routing capability is limited (as was formerly often the case), all calls from directly connected users may be routed to the area's tandem exchange, which then routes the local calls back to the local exchange from which they began. The call then ends up being routed through both the local and tandem exchanges (unless the local exchange is smart enough to spot what's happening and *drop back* the call). The looped-back shape of the call is allegedly like a trombone; in the case of Figure 7.10, one played by a recumbent trombonist. Now, who says network routing isn't interesting?

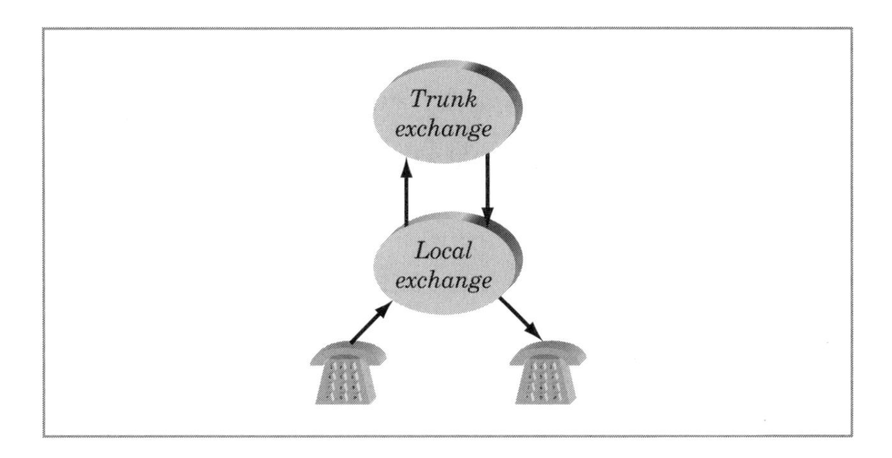

Figure 7.10 Trombone working

Automatic alternative routing

Automatic alternative routing (AAR), illustrated in Figure 7.11, is a way of achieving efficient use of direct routes, while using the backbone network for load peaks.

The direct (also known as *high usage*) routes between local exchanges are deliberately under-provided, so that they are kept well occupied by normal traffic. The switches are programmed to route preferentially over the direct routes (shown as 1), but to fall back to routing via the tandem (or *final*) routes (shown as 2) when the direct routes are fully occupied.

Crank back

Where SS7 common channel signalling is implemented (and that means pretty well everywhere), the rich information provided by SS7 can be used to support more complex routing heuristics, such as *crank back* (also, confusingly, called *drop back*), as illustrated in Figure 7.12. In this example, a user on switch A tries to call a user on switch B. Switch A's direct route to B is fully occupied, so switch A (by AAR or otherwise) passes the call to switch C. Although there is free capacity between A and C, C finds that it has no free capacity on its trunk to B. C reports this to A, which then tries (successfully) to route via D instead.

Dynamic routing

Dynamic routing, shown in Figure 7.13, means changing routes over time, which is particularly relevant when a network spans multiple time zones. The classic example is from AT&T's long-

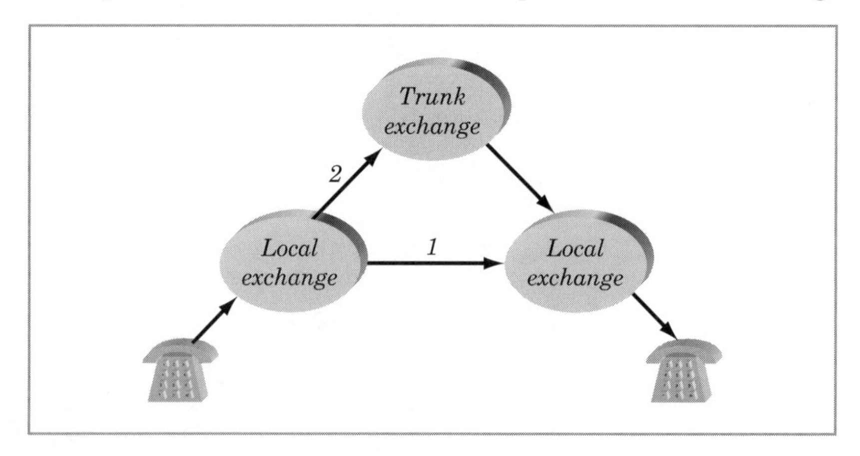

Figure 7.11 Automatic alternative routing

distance network in the US. Here, there are trunks between San Francisco (in the west of the continent) and both New York and Washington (in the east). Late in the day (eastern time), those trunks are busy. But early in the day, when the users in the east are at their busiest, the folks in San Francisco are asleep, and those east–west trunks are scarcely used. Taking advantage of this, the direct route between New York and Washington is under-provided. The East Coast switches are programmed to route New York–Washington

The East Coast switches are programmed to route New York–Washington calls via San Francisco *in the morning.*

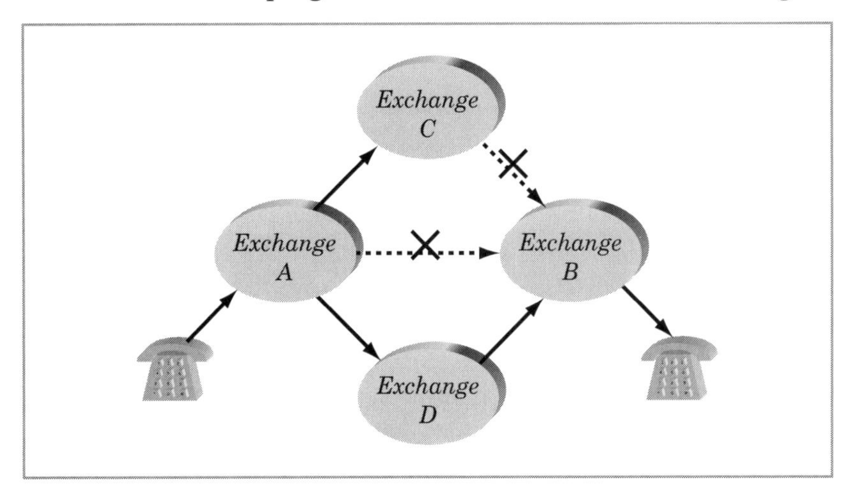

Figure 7.12 Crank back

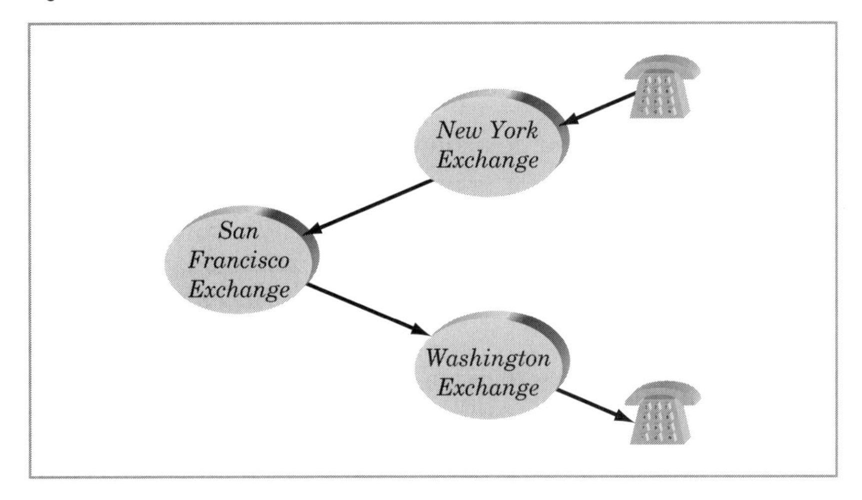

Figure 7.13 Dynamic routing

calls via San Francisco in the morning (eastern time), but to stop doing that later in the day when there will be genuine east–west traffic to consider.

Real-time network routing

AT&T has developed a second-order version of AAR, called *real-time network routing* (RTNR). Each switch monitors the load not only on all its direct connections but also (in collaboration with its neighbours) on its two-link routes (like a chess player looking two moves ahead), and so routes calls on the basis of the best two-link option.

Trunk reservation

The automatic use of two-link routes for excess load is not without problems. It tends to use the network less and less efficiently as the offered load increases, and can mean that under high offered load the network traffic actually decreases. To limit this problem, *trunk reservation* marks some circuits from a route as for first choice use only, and not for use as part of an overflow route.

Number portability and routing

Increasingly, network operators are being required to allow their customers to move (or *churn*) to other operators, but to take their numbers with them. This has been a major technical challenge for many operators, whose networks were designed around the expectation that all numbers, for example, with a given area code, would be in one network.

Number porting can be achieved through enhanced routing features, as follows:

- the identity of the destination exchange (the one the customer has moved to) is recorded in the routing tables of the donor exchange (the one the customer has moved from);
- when a call to the ported number is made, it arrives at the donor exchange, which receives the call and passes it back into the (SS7) network with a prefix to the called number, identifying the destination exchange;
- if the call came in from the trunk network, then either the call trombones back to trunk exchange (which

unnecessarily busies two extra trunks), or, if the trunk network is smart enough, it drops the trombone link and makes the connection directly.

There are also *intelligent networks*, solutions, where each call is screened on entry to the network, to check whether the called number has been ported, and to redirect the call if necessary. While these reduce the load on the network, they require very substantial processing capacity in the IN equipment, which has to screen every call against a possibly very long list of ported numbers.

Intelligent networks (IN)

Why IN?

Intelligent networks is a circuit switching technology which is typically overlaid onto a conventional network as an enhancement. The motivation for the development of IN was the telcos' desire to be able to offer new and differentiating network services, and to get them to market quickly. Ordinary SPC circuit switching technology would not do, for a number of reasons.

Firstly, for a telco to get a major change made to the software of its SPC switches could take anything from two to five years. The software in large SPC switches is typically bogglingly complex and intransigent. Large, mature SPC products are very serious things to modify. And then there are the switch vendors' development programmes to contend with. A switch vendor will want to satisfy a broad market, not just one telco. So, if a telco asks for a new feature, it may have to wait a while before the switch vendor will fit it in to its product development programme.

Secondly, upgrading the software build of all the switches in a large network is very costly because of the very careful testing that has to be done, and simply because of the mechanics of loading and configuring an immense new software build onto thousands of machines across a wide area. Also, it is disruptive, and may necessitate switch downtime. Worst of all, it is risky. No matter how hard telcos try to test out new switch software builds in the labs, there are still serious risks of real network conditions exposing an unforeseen problem, and quite possibly crashing the network. In such an event, rolling back the network software build can take days, and can seriously damage a telco's market presence.

Thirdly, many desirable advanced network services require the co-ordination of several switching nodes, which may in many networks be from more than one supplier. Co-ordinating the development programmes of competing suppliers is generally prohibitively difficult.

From POTS to PANS

The core idea of IN is to separate the *service logic* of the network from the *switching*.

IN, then, is a technology which was trumpeted as being able to take telcos from offering POTS (*plain old telephony service*) to PANS (*pretty awesome*[8] *new stuff*). The core idea of IN is to separate the *service logic* of the network from the *switching*, by moving the service logic onto general-purpose computers. The service logic programs can then be modified, without recourse to the switch vendors, to develop new network features and services quickly (and possibly even cheaply).

The kind of network services included in PANS include:

- freephone (either to a single destination number, or to the nearest of several destinations) and similarly premium rate access;
- charge card services, including some forms of prepaid mobile service;
- network ACD (with proportional distribution across multiple sites, or to the least loaded);
- personal numbering and number portability;
- call-completion services (services designed to cause calls to get through to something or other, even when the proper destination is busy, so that the telco can bill for the calls), such as divert on busy or on no reply, or network voice mail;
- universal access numbers for specialized services (e.g. 192 for directory inquiries).

[8] Or, in the UK, *amazing*.

IN architecture

The principal elements of an IN solution, as defined in ITU-T recommendations Q.1200–Q.1290, are shown in Figure 7.14, and described in the following paragraphs.

Switching the calls

The *service switching point* (SSP) is a modified version of (or adjunct to) an ordinary SPC switch. The modifications are as follows. The SSP looks out for a range of trigger events, such as dialled digits, call attempt failure or call completion. When a trigger occurs, the SSP temporarily pauses the progressing of the call, and refers the call to the SCP (*service control point*), supplying the SCP with information about (at least) the called and calling addresses, and the trigger event.

The SSP also offers to perform, on command, a range of elementary switching functions, including:

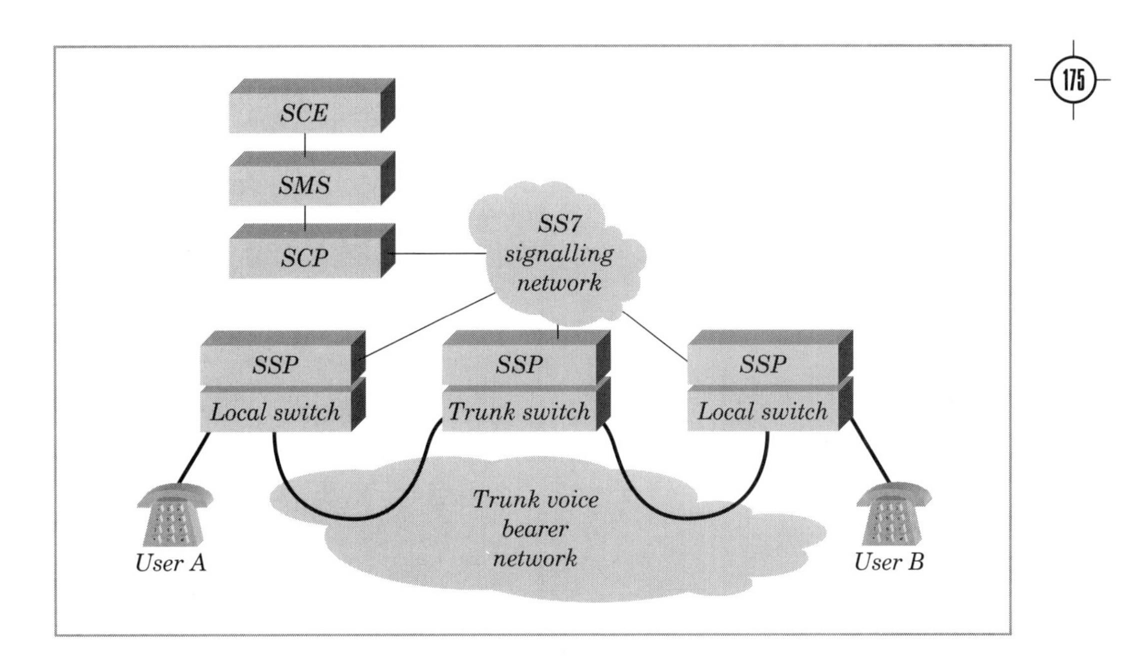

Figure 7.14 IN architecture

- connect call to a specified destination;
- play an announcement;
- release the call, and so on.

The ITU-T has defined a basic repertoire of SSP behaviours, called *capability set 1* (CS-1), and defined in ITU-T recommendation Q.1210–Q.1219. Obviously, to have an IN network, you first have to persuade your switch vendor to implement at least CS-1. This may be easier, however, than getting it to implement special custom functions just for you.

Call control

In an IN network, call control is handed over from the switches to one or several *service control points*. The constituents of an SCP are outlined in Figure 7.15.

An SCP hosts one or several *service logic programs* (SLPs). Each IN service offered by the network will have its own SLP. Each SLP will be constructed out of programming elements called *service independent building blocks* (SIBs). Examples of SIBs are 'translate', 'verify', 'compare' and 'queue call'. The software environment that the SIBs require is called a *service logic execution environment* (SLEE), which will in turn rely on *SCP node software* (low-level software elements such as a database and an SS7 stack). The SCP will typically be implemented as a very high-performance, high-availability (often fault-tolerant) computer.

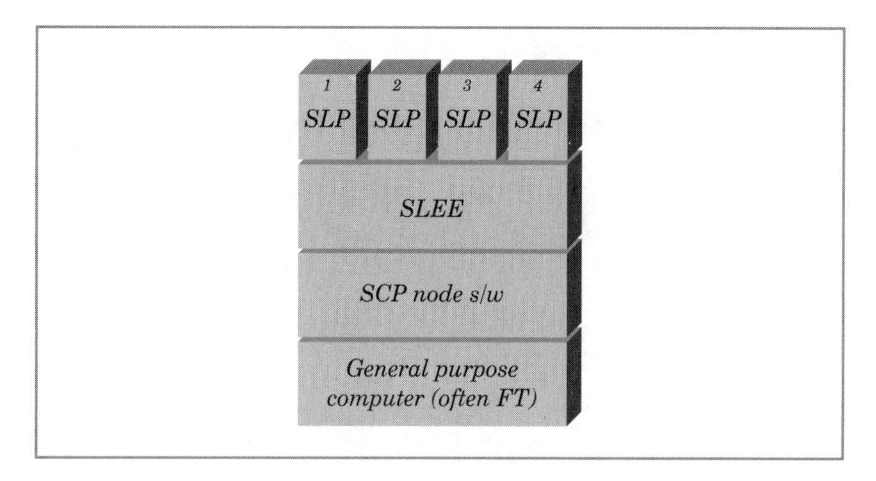

Figure 7.15 Service control point

The signalling network infrastructure

The IN components are connected over an SS7 network, including at least SCCP and TCAP, and an *IN application part* (INAP) on top of TCAP.

Writing a service logic program

Service logic programs are constructed offline, in a *service creation environment* (SCE). An SCE allows users to construct service definitions out of SIBs, often by moving them around, adjusting their parameters and connecting them, in a service flowchart on a slick and glossy user interface. An SCE also serves as a tool for requirements capture and service specification, often providing facilities for prototyping and demonstration, so that a network operator can work closely with semi-technical customer staff to agree an appropriate service.

A network operator can work closely with semi-technical customer staff to agree an appropriate service.

From the back room to the live network

A *service management system* (SMS) supports the IN network by distributing released SLPs to the SCPs. It also distributes provisioning data. For example, there may be a single generic SLP for simple 0800 number translation. Every time the network operator sells its 0800 service, the IN network has to be provisioned with, at least, the new 0800 number and the corresponding destination telephone number.

The SMS may also check data consistency between the several SCPs in the network, and may include quite challenging software for rolling back the state of all the SCPs in the network, if a complex update fails.

IT-U intelligent networks recommendations

Q.1201/I.312 Principles of intelligent network architecture (the same recommendation published under two identities)

Q.1202/I.328 Intelligent network–service plane architecture.

Q.1203/I.329 Intelligent network–global functional plane architecture.

Q.1204 Intelligent network distributed functional plane architecture.

Q.1205 Intelligent network physical plane architecture.

Q.1208 General aspects of the intelligent network application protocol.

Q.1211 Introduction to intelligent network capability set 1.

Q.1213 Global functional plane for intelligent network CS-1.

Q.1214 Distributed functional plane for intelligent network CS-1.

Q.1215 Physical plane for intelligent network CS-1.

Q.1218 Interface recommendation for intelligent network CS-1.

Q.1219 Intelligent network user's guide for capability set 1.

Q.1290 Glossary of terms used in the definition of intelligent networks.

Intelligent peripherals

Often closely associated with IN networks are devices called *intelligent peripherals*, which are occasionally, confusingly, abbreviated to IP. Intelligent peripherals provide auxiliary services such as voice guidance, digit collection, voice recognition and mailbox facilities.

Service nodes

What I have described so far is the classic IN architecture presented in the ITU-T specifications. In real life, things are often not so neat. Often an SCP will not support SLPs defined as SIBs, but will require the SLPs to be written in a programming environment which is specific to the SCP. Also, the SCP is often not fully separated from the switch. Instead, vendors offer small switches, with built-in SCP functions, called *service nodes*. While these do not offer all the freedom of a classic IN architecture, they offer an attractive mix of flexibility, economy and availability.

IN, but not as we know it

The ITU-T IN standards are not the only ones. Telecordia (Bellcore) has a set of standards for *advanced intelligent networks* (AIN). AIN is, however, so similar to ITU-IN (and not obviously any more advanced) as not to warrant separate discussion here.

Another alternative to classic IN is JAIN (Java in IN). The idea is that lots of SS7 stack vendors should provide Java programming interfaces for their stacks, and that IN programs are written in Java, using Java beans in place of SIBs. The role of the SCE is taken over by a bean development kit, and the *Java dynamic management kit* (JDMK) offers ready-made network management facilities over SNMP (*simple network management protocol*) or HTML. By using Java, extreme portability is assured, as is integration with the rising tide of IP applications.

While there are some technical arguments about the relative value of classic IN, JAIN, and other emerging alternatives such as *Parlay*, the most serious doubt for their future comes from the pre-existence of classic IN, and the ascendancy of the Nethead point of view described below. There simply may not be enough room, in a possibly shrinking IN market, for another set of standards.

Or the end of IN?

There are reasons for suspecting that IN has passed its golden age and will become marginalized. This possibility is the subject of the celebrated difference of opinion between the 'Bellheads' (the network operators) and the 'Netheads' (the Internet user community). The Nethead argument goes roughly like this:

- IN solutions, by embedding the intelligence within the network, use the network's connectivity efficiently. But now connectivity is nearly free, efficiency in its use is no longer important.
- The 'solutions' offered by IN have failed to liberate the telcos. Even IN services still take a long time to develop, test and roll out. Also, IN solutions are rarely transparent to the network switch type, so that they do not allow the network operator to mix and match switches (or switch vendors).

- IN is perceived as an attempt by the network operators to keep control of the technology of offering value-added services.
- It would be better, say the Netheads, to put the intelligence outside the network, and give control of services directly to the application providers.
- Solutions could be built using Internet protocols and large numbers of cheap computers, rather than SS7 and huge, powerful, fault-tolerant computers.

It appears very much that things are moving in the direction of the Nethead point of view.

Further reading

For most of the material in this chapter, Flood (1997) provides the next level of detail. Flood's European perspective may lead US readers to choose Cole (1999) as a US-oriented supplement. Cole also has a lot to say (to all readers) about PABXs, ACD systems and their management. The Nethead/Bellhead argument is presented in Isenberg (1998).

8 Packet mode subnetwork technologies

Packet mode networks are built out of several alternative or complementary subnetwork technologies. Most packet mode networks consist of a number of *subnetworks*. For example, there may be a LAN at the near end, a WAN (*wide area network*) in the middle, then another LAN at the far end. This chapter looks at the technologies that are used for constructing such subnetworks. The next chapter then looks at the glue that assembles subnetworks, to provide end-to-end services.

Connections and packets

Every packet mode technology is either connection oriented or connectionless. In a connection-oriented network, user A asks the network for a virtual connection (or virtual circuit) to user B; to do this, A has to quote B's address in full. The network finds a route to user B, and gives both A and B their own reference labels for the connection. Then when A wants to send data to B, A only has to quote the right label, not B's full address (and vice versa when B wants to send to A). There are no network resources committed to the connection, however, apart from a few label-to-route mappings. Every packet of data that is sent has to contend for use of the network. A connection-oriented packet network has the advantages of simplicity (once the connection is established) and of having a context within which successive packets can be sequenced, so that the integrity of a multi-packet message can be established. It achieves this at the cost of some complexity in the call establishment phase.

Not all virtual circuits are overtly established in this way.

The approach described above is for a *switched virtual circuit*. Where two parties are expected to have a regular need for communication, the network can be taught (by a management system) to maintain a permanently open connection, or *permanent virtual circuit*.

The alternative to all this is to have a connectionless network. In such a network, A can send to B at any time, without having to ask for a connection first. A includes B's full address (as well as A's own) in each packet that is sent. A packet like that is called a datagram. The network routes each datagram separately, and has no concept of an ongoing connection between A and B. While this is in many ways simpler, it loses out on the contextual structure that a connection-oriented network offers. Also, it incurs the overhead of transmitting both parties' full addresses over and over again.

All packet networks differ fundamentally from circuit mode networks in that:

- they present a variable end-to-end transfer delay, as each packet may follow a different path through the network,[1] or have to be queued within network nodes while they wait for spare link capacity;
- the network can block the data stream even after some data has got through;
- packets can get lost in the network, or can arrive in the wrong sequence, or can arrive duplicated.

We will see in this and the next chapter how packet network technologies attempt to overcome these limitations.

Telex

The earliest data networks were circuit mode telegraph networks, and it was from these that *telex*, the first packet mode network technology, was derived. Telex is strictly speaking a message switching system, not packet switching. That is to say, the whole message is sent in one chunk, rather than chopping it up into little packets.

[1] This is theoretically the case, but in real life it is unusual.

A telex user terminal uses a permanent physical circuit to a local telex switch. The user dials the network address of the destination and the network establishes a virtual connection. The user terminal then sends the message using a 50 bps modem, and the ITA (*international telegraphy alphabet*) Number 2 code, a direct descendant of the Baudot 5-bit character code used in telegraphy systems. This is passed through a network of telex switches and then delivered over a similar interface at the far end.

The physical connections out to the user terminals will carry only one connection at a time. However, within the telex network, messages are multiplexed so that a network link will carry several different users' messages in swift succession.

Telex networks are still operated, but are obsolescent because of their low speed and poor utilization of the local physical connection.

X.25

In the 1960s, computers became cheap and common enough[2] to warrant connecting computer systems together into packet switching networks. The packet switching standard that emerged was ITU-T recommendation X.25.

X.25 was based on the assumption that the links between the network nodes would be error prone. The telco bearer networks that provided the links were then all analogue, and the error rates for modems in those days were much higher than they are now. If a packet was allowed to traverse a network unchecked, across several links in turn, the chance of it reaching the far end without error was poor. Therefore X.25 specified that each link in the network should operate its own error checking and retransmission mechanism. This means that every node along a packet's path has to run the packet through a lot of software, which in turn limits network throughput.

Recommendation X.25 only really addresses the network-to-user interface. An X.25 network can use any protocols it likes, inside. Usually, an X.25 network will use a derivative of X.25 for its internal links. At the edge of the

An X.25 network can use any protocols it likes, inside.

[2] At least in the military and academic circles from which X.25 originated.

network, the user equipment is designated the *data terminal equipment* (DTE) and the network equipment is designated the *data communications equipment* (DCE). The interface is defined as a stack of three levels:[3]

1. The physical interface (V.28 electrical signals and V.24 signalling convention).
2. The *link access protocol*, which is a form of '*high-level data link control*'. Two options for the LAP are allowed, a symmetrical one called LAP-B (for *balanced*) and an asymmetrical one called LAP.
3. The *packet level protocol* (PLP), which looks after the routing of packets, the management of PVCs and SVCs, network flow control and end-to-end packet sequencing.

Between X.25 networks, another interface standard, X.75, is used. X.75 is in most respects a simplified version of the X.25 packet level protocol.

While X.25 was useful for communication between computers, it did nothing to address the requirements of character mode terminals. Often a number of co-located terminals would need to access one or more mainframe computers in distant data centres. X.25 was seen as a way to multiplex several terminals' sparse traffic onto a small number of links.

To support this, ITU-T recommendations X.3, X.29 and X.28 (collectively called 'triple-X') defined how a *packet assembler and disassembler* (PAD) should behave.

As well as military and academic X.25 networks such as ARPANET and JANET (*Joint Academic Network*), X.25 networks were (and still are) offered commercially by most telcos. X.25 was the network layer foundation of the *open systems interconnection* (OSI) standard protocol stack, even though X.25 was not in itself quite conformant to the OSI view of what a network layer protocol should achieve. The OSI stack was mandated noisily by various institutions during the 1980s but has now been marginalized by the overwhelming success of IP networks. OSI protocols still pop up from time to time (for example, FTAM, a *file transfer and manipulation protocol* functionally similar to FTP, and X.400, a messaging protocol

[3] *Levels*, not *layers* as in OSI.

comparable to SMTP, but they are rapidly diminishing in significance.

Frame relay

As telco bearer networks digitized in the 1970s and 1980s, data-link speed and quality increased. X.25 networks began to look slow and processor-bound. The Frame Relay Forum developed standards for an alternative packet mode technology, based on the new realities of telco networks.

Frame relay, like X.25, offers a connection-oriented service. Like X.25, it ships the users' data packets across the network in containers called frames. However, it exploits low link error rates by only attempting to check for lost frames, and request their retransmission, end-to-end across the network. On each data link, there is error checking, but errored frames are just thrown away by the intermediate nodes.

Unlike X.25, there is enough information visible at the frame level for intermediate nodes to route frames without having to interpret their contents at a higher level, so the packet level protocol executes only in the network edge devices.

Frame relay also attempts to offer management of *quality of service* (QoS). In an X.25 network, all packets and all virtual circuits have equal priority, and so are all equally subject to delay or loss if there is congestion. Frame Relay goes one better: it allows a user to request a *committed information rate* (CIR) when setting up an SVC (or for the management system to request one when defining a PVC). The network attempts to limit its commitment to the capacity that it can offer. If a user sends data faster than the negotiated CIR, the network node at that user's point of connection marks the excessive frames as eligible to be discarded within the network if congestion arises.

The ITU-T recommendations for frame relay begin with I.122, the *framework for frame mode bearer services*. Frame relay services are now offered by most telcos, at data rates from 56 kbps up to 45 Mbps.

SMDS

Another high-speed packet-oriented subnetwork technology is the *switched multimegabit data service* (SMDS), also known as the *connectionless broadband data service* (CBDS). SMDS offers a connectionless service, with packet sizes up to 9183 octets.[4] The network-edge interface offered by SMDS is called the *SMDS interface protocol* (SIP), specified in Telecordia standards TR-TSV-000772 to 775, and operating at rates up to 34 Mbps.

Broadband integrated services networks

We have seen how the N-ISDN attempts to offer a variety of teleservices (voice, video, data, etc.) over a single interface. However, N-ISDN interfaces do not go fast enough to support the demands of many modern applications. For example, colour video conferencing requires up to 2 Mbps; studio picture distribution requires 135 Mbps; CAD systems need to transfer volumes of data that would be impractical over an N-ISDN interface.

While these high data rates can be achieved over specialized, committed interfaces, many users would like to have a single network access technology that could carry all of their traffic. There thus appears to be a market demand for something like the ISDN, but faster. Anything going at 2 Mbps or above is called 'broadband', and so the idea of the *broadband ISDN* (B-ISDN) has emerged.

The demands on such a B-ISDN technology are challenging. For example:

- users want their network interface to have enormous capacity when they want it ('bandwidth' on demand), but not to have to pay for that when they aren't using it;
- their data traffic, while often tolerant of delay, requires very low error rates (called *semantic transparency*);
- their speech traffic, while tolerant of quite high error rates, requires strict control of delay and jitter (*time transparency*).

[4] While 'octet' does have its own rather technical meaning, for most purposes it can be taken as a fancy word for 'byte'.

I introduce the concept of the B-ISDN here because ATM, the next technology discussed, was conceived as the solution to the B-ISDN problem. Indeed, B-ISDN and ATM are still sometimes used quasi-synonymously. However, ATM has for various reasons not become the full answer to the B-ISDN problem. As we will see in the next chapter, IPv6 and other IP technologies may also have much to contribute to the B-ISDN.

ATM

Asynchronous transfer mode (ATM) was designed[5] to offer a single network interface for all kinds of network traffic, for example:

- LAN interconnection;
- video distribution;
- Web access;
- remote imaging;
- multimedia;
- voice services.

ATM therefore offers a range of connection-oriented services, including:

1. *Constant (or continuous) bit rate* (CBR) for traffic which will be offered to the network at a steady rate, such as non-compressed video. A CBS connection is specified in terms of its required *peak cell rate* (PCR).
2. *Variable bit rate* (VBR), for traffic which will be offered to the network at a variable rate. A VBR connection is specified in terms of PCR, *sustainable cell rate* (SCR), and *maximum burst size* (MBS). VBR services include:
   - real-time VBR (VBR-rt), with constraints on both the absolute delay introduced by the network and the delay variability;

[5] Under the auspices of the ATM Forum, which defined standards subsequently adopted by the ITU-T.

- *non-real-time* VBR (VBR-nrt), without such delay constraints.

3 *Available bit rate* (ABR) for traffic sources which are elastic and which can guarantee to use whatever capacity the network offers from moment to moment.

4 *Unspecified bit rate* (UBR) where the network reserves no capacity for the service, and the service just gets to use whatever capacity is available. UBR services are appropriate, for example, for file transfer and email.

The ATM protocol architecture is designed to keep the devices within the ATM network simple.

The ATM protocol architecture is designed to keep the devices within the ATM network simple, so that they can be implemented without software involvement in the data streams. As Figure 8.1 shows, the network nodes are required only to support the two lower layers of the ATM stack. The *network/network interface* (NNI) between ATM nodes is kept largely ignorant of the nature of the traffic passing through it, and knows just enough about its service requirements to maintain the appropriate QoS.

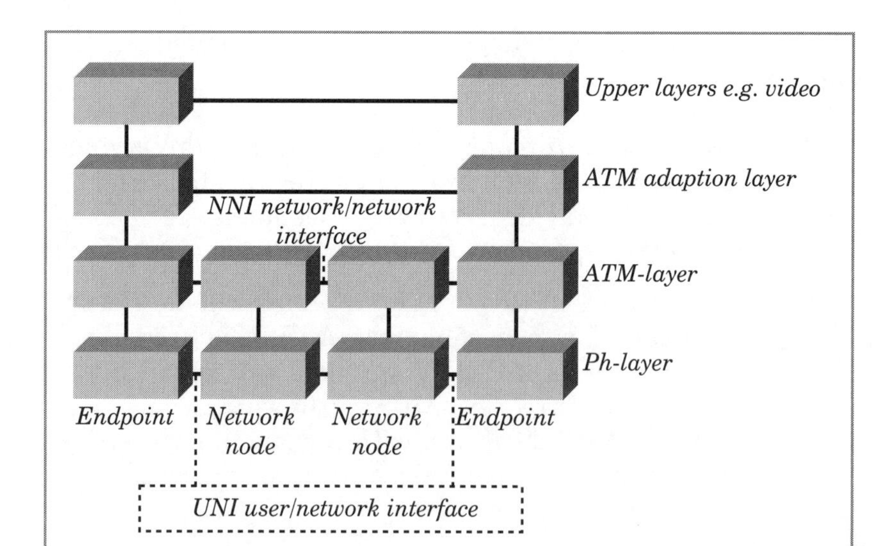

Figure 8.1 ATM protocol architecture

The ATM adaption layer and higher-level protocols are implemented only at the edges of the network. They are divided into user features (the U plane), signalling control (the C plane) and network management (the M plane).

Physical layer

In an effort to become a universal service delivery vehicle, ATM offers a wide range of physical layer interfaces based on both LAN and WAN technology (including Sonet, SDH, T-1/E-1 PDH and Category 5 twisted pair cabling).

ATM layer

The distinctive cleverness of ATM comes in the 'ATM layer'. This has to achieve a compromise between the demands of the very different service requirements that ATM has to support. The basis of the ATM layer is cell switching. The cell structure and the associated protocols are independent of the physical medium. Rather than switch packets or frames, which are of variable length, ATM networks switch cells, which are fixed length and therefore amenable to simple hardware designs. Each ATM cell is 53 octets long. This is a compromise between the inefficiency of very small cells, where the inevitable header information would take up a relatively large part of the cell, and the unacceptable switching latency and transit delays which large cells would introduce.

Because an ATM channel can carry cells relating to several connections, a newly arrived cell may have to wait a while in an ATM switch for a free cell space on the channel. Therefore, a connection's ATM cells do not arrive at exactly regular intervals, which is why it is called *asynchronous* transfer mode. Each ATM cell consists of a 5-octet header followed by a 48-octet payload. The content of the payload is the business of the higher protocol layers, and is just carried transparently by the ATM layer. The header includes routing information indicating which virtual connection it belongs to, and also (importantly) a cell-loss priority bit, indicating whether it can be discarded if congestion occurs. The header has an error detection checksum, so that the integrity of the routing information is assured, but the payload does not (payload integrity has to be handled in higher layers).

The primary job of an ATM network node is cell switching:

receiving cells from one port and spitting them out at other ports, depending on the routing information in their headers. The simplicity of the ATM header means that ATM switches can select the outgoing port in hardware, and therefore can run very fast.

ATM switches

ATM switch architectures are complex, and the following examples are much simpler than most real switches.

Figure 8.2 illustrates a *time-division switch*, where a high-speed backplane collects all the incoming cells, and then the output ports are served by controllers which pick off the cells they want (according to routing tables implemented in hardware).

Figure 8.3 illustrates a *space-division switch*, where a cross-point matrix provides an alternative way to connect all the inputs to all the outputs.

A third basic architecture for ATM switching is the *self-routing* or *delta switch*, shown in Figure 8.4. Here the routing function is distributed across a matrix of simple switch stages. When a cell arrives, the input controller consults the switch

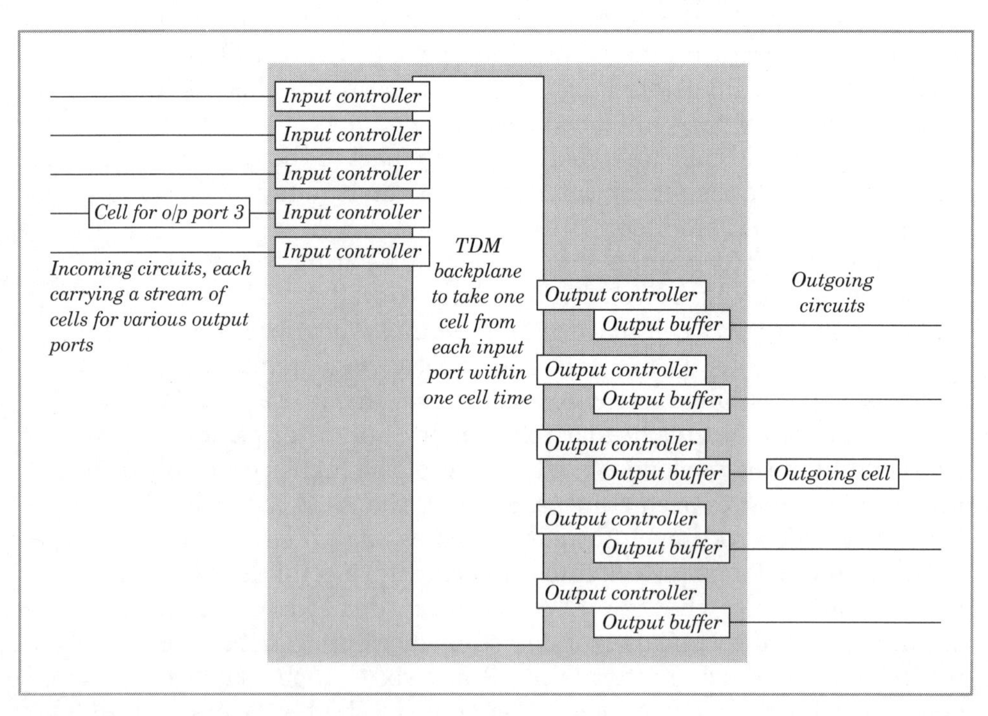

Figure 8.2 Time-division ATM switch element

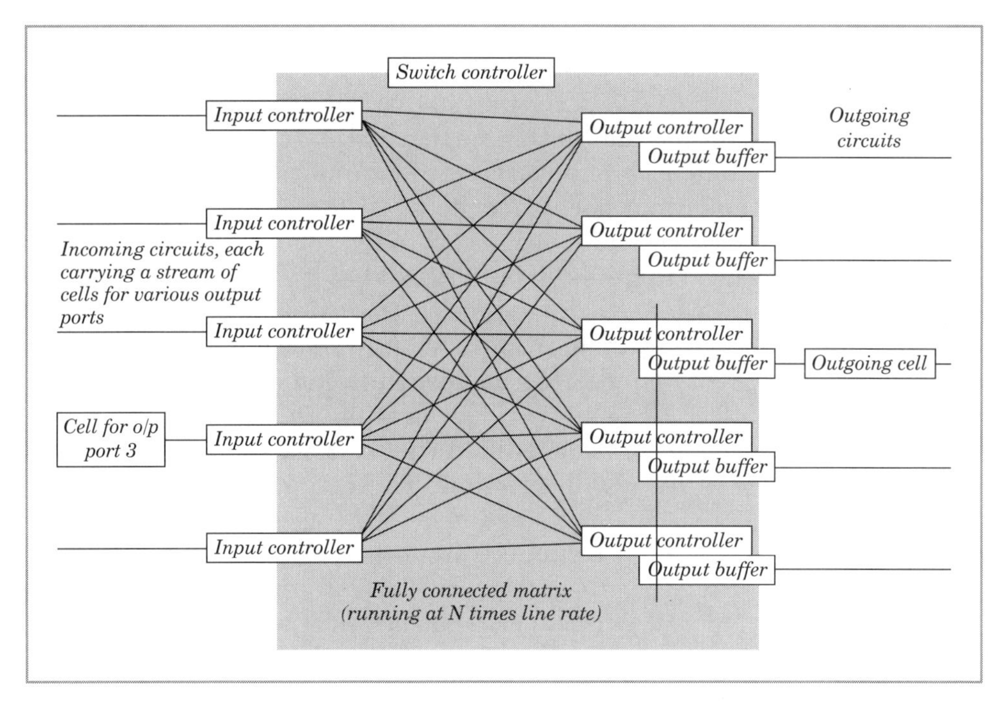

Figure 8.3 Space-division ATM switch element

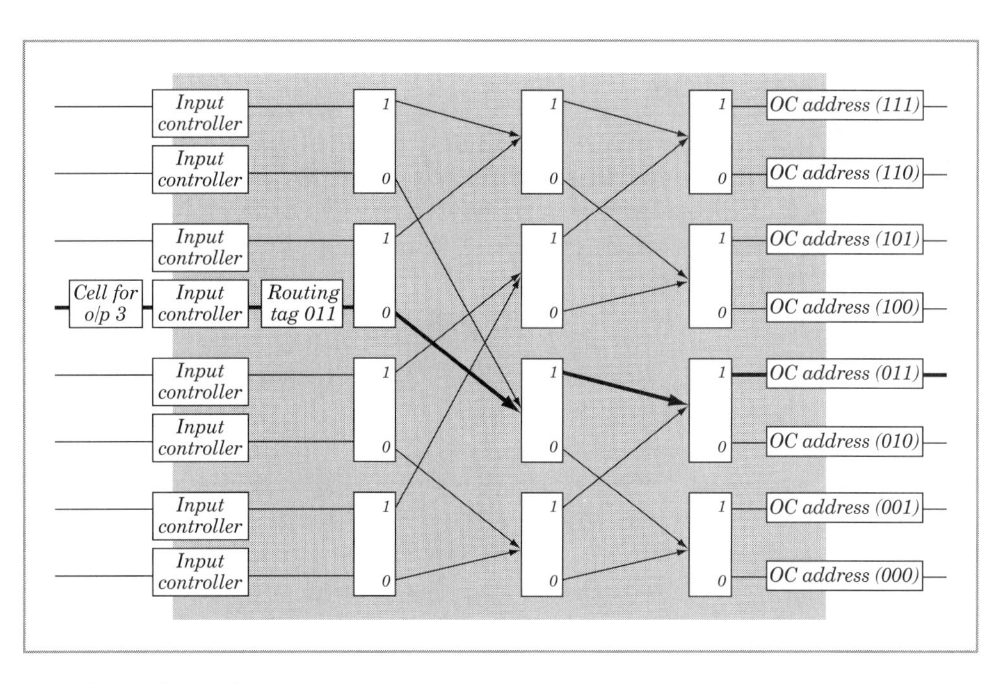

Figure 8.4 Delta switch

controller and determines which output port it should go to, and prefixes the cell with a routing tag identifying the output port. Then each switch stage in turn interprets successive parts of the routing tag (in this small example, the switch stages are binary and each one interprets just one bit of the routing tag) until the output port is reached.

In all cases, the output controllers or intermediate switch stages need to have cell buffers, so that if there is not a free cell space on the outgoing channel, the cell can be queued. Real switches contain multiple switching stages, to achieve the desired balance between blocking probability and cost. However big the output buffers are, if the offered load is large enough, the buffers will overflow. To ensure that QoS commitments are met, ATM uses the 'leaky bucket' technique (properly called the *generic cell rate algorithm* or GCRA) to dump cells belonging to less critical classes of service when queues overflow. The GCRA measures the traffic volumes on virtual connections, and allows cells which exceed the contracted rate to be marked as eligible to be discarded if congestion occurs.

Real switches contain multiple switching stages, to achieve the desired balance beteen blocking probability and cost.

Channel structure

As Figure 8.5 shows, ATM has quite an elaborate channel structure. A physical link can carry several *virtual paths* (VPs), and each VP can carry several *virtual channels* (VCs). Routes can be requested at the level of an individual VC or, where several

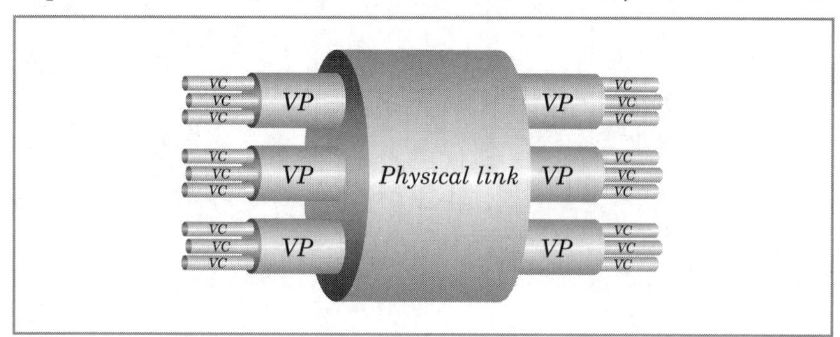

Figure 8.5 ATM channel structure

connections are required between the same network end points, at the VP level.

ATM adaption layer

The business of the ATM adaption layer (AAL) is to use the facilities of the ATM layer to deliver a range of PVC and SVC services. Four standard classes of service are defined:

- AAL 1 (alias Class A) offers a connection-oriented low-latency CBR service, with no guarantee of delivery. AAL 1 is used, for example, for voice streams that have no silence suppression;
- AAL 2 (alias Class B) is a connection-oriented delay-sensitive VBR service, used for example for voice streams which do have silence suppression. Both AAL 1 and AAL 2 maintain the timing relationship between the ends of the connection. The other classes following do not;
- AAL 3/4 (Class C/D) began life as two distinct classes which have since merged. AAL 3/4 and AAL 5 both provide connectionless VBR services, for example for LAN interconnect, for carrying SMDS or for frame relay;
- AAL 5 is a cleaner, simpler alternative to AAL 3/4 (it leaves out the facility for re-sequencing within a VC), and is therefore called the *simple and efficient adaptation layer* (SEAL). AAL 5 is also used for the ATM C plane, where it is called the *signalling AAL* (SAAL).

AAL is divided in the standards into two sublayers. The upper one, the *convergence sublayer* (CS) maps the AAL service onto the service offered by the ATM layer. The *segmentation and reassembly* (SAR) sublayer packs the carried *protocol data unites* (PDUs) into cell payloads, and reassembles them at the far end.

ATM signalling

The UNI signalling is a variant of DSS 1, in which DSS 1's Q.921 data-link layer is replaced by the faster *service-specific connection-oriented protocol* (SSCOP). Also DSS 1's Q.931 network layer protocol is extended for broadband, in recommendation Q.2931.

LAN emulation

The ATM *LAN emulation* (LANE) protocol supports the transparent interconnection of Ethernet, Token Ring and FDDI LANs.

Market penetration of ATM

The inventors of ATM hoped that it would become a universal network technology. However, things have not worked out that way.

In LANs, Ethernet has increased its speed by a factor of 100, giving it performance comparable to ATM. The large market for Ethernet equipment has kept the cost low, so Ethernet is a much cheaper alternative in the LAN.

In telco WANs, ATM has made considerable progress, but even there it is not unchallenged. The ATM fixed cell size can lead to inefficiency in channel usage. For example, a 40-byte message with an 8-byte checksum just fills an ATM cell's 48 octets of payload space, and so the inefficiency due to the ATM header is only 9 per cent. But a 41-byte message would require two cells to carry it, and so raise the overhead (sometimes called *cell tax*) to 61 per cent.

The position of ATM in network architectures is discussed further in the next chapter.

ITU-T ATM recommendations

I.361	B-ISDN ATM layer specification
I.362	B-ISDN ATM adaptation layer (AAL) functional description
I.363	B-ISDN ATM adaptation layer specification
I.363.1	Types 1 and 2 AAL
I.363.3	Types 3/4 AAL
I.363.5	Type 5 AAL
I.364	Support of the broadband connectionless data bearer service by the B-ISDN
I.365	B-ISDN ATM adaptation layer sublayers
I.731	Types and general characteristics of ATM equipment
I.732	Functional characteristics of ATM equipment
I.751	Asynchronous transfer mode management of the network element view
G.804	ATM cell mapping into plesiochronous digital hierarchy (PDH)
Q.2100	B-ISDN signalling ATM adaptation layer (SAAL) overview description
Q.2110	B-ISDN signalling ATM adaptation layer – service-specific, connection-oriented protocol (SSCOP)
Q.2119	B-ISDN ATM adaptation layer protocols – convergence function for service-specific, connection-oriented protocol above the frame relay core service
Q.2120	B-ISDN meta-signalling protocol
Q.2130	B-ISDN signalling ATM adaptation layer – service-specific co-ordination function for support of signalling at the user–network interface (SSCF at UNI)
Q.2140	B-ISDN signalling ATM adaptation layer – service-specific co-ordination function for signalling at the network node interface (SSCF at NNI)
Q.2144	B-ISDN signalling ATM adaptation layer – layer management for the SAAL at the network node interface

Point-to-point protocol

The point-to-point protocol originated as a data link protocol to connect PCs over modem links to their Internet server.

The *point-to-point protocol*, or PPP, defined in RFCs 1548, 1661 and 1662, originated as a data link protocol to connect PCs over modem links to their Internet server. It is therefore a point-to-point full duplex serial data-link protocol, using HDLC-style framing and link control, over either asynchronous or synchronous physical layer options.

PPP is used not only to carry IP but also proprietary protocols such as SNA, Vines, DECnet and NetBIOS. The core protocol is supported by a number of ancillary protocols, including a *link control protocol* (LCP, defined in RFCs 1570 and 1661) for establishing, configuring and testing the data-link connection, and *link quality report* (LQR, defined in RFC 1333) which reports link statistics.

An extension of PPP, PPP multilink (RFCs 1717 and 1990), allows multiple (parallel) physical links to be combined into one logical link, with facilities for segmenting the carried PDUs into pieces for transmission across the links, and then reassembling them at the other side.

As well as its original use, PPP is used in long-haul networks, on top of SDH or Sonet, as an alternative data-link layer to ATM. Confusingly, this usage is often called IP over Sonet/SDH, without bothering to mention the PPP. PPP over Sonet/SDH is defined in RFC 1619.

Further reading

For a general introduction to packet mode communications, it is difficult to beat Halsall (1995). To address the detailed technicalities of ATM, Goralski (1995) is a good supplement.

9 Internet technologies

The *internet protocol* (IP) suite is by far the fastest growing set of protocols at the network layer and above, supplanting OSI, SNA, and DECnet for example. This growth is both in market presence, and in the sheer number of protocols defined. This burgeoning of protocols is because of the large number of interested parties available to develop them, the rapid development of new internet applications, and a 'suck it and see' attitude, where often several protocols will be developed to address a given problem, and it will be left to the market to select the fittest.

A consequence of this is that it is not practical to describe all the internet protocols usefully in one book, let alone in one chapter. This chapter therefore confines itself to the best-known and most widely used internet protocols, as presented in Figure 9.1.

Internet network layer protocols

Internet protocol

IP, the internet protocol, is intended for routing datagrams between subnetworks, without concern for what the subnetwork technologies are. IP is a connectionless protocol, and therefore the devices that operate it do not have to remember lots of information about the status of connections. This simplicity contributes to the success and scalability of IP.

IP comes in two flavours. The predominant flavour is version 4 (IPv4), which has been around for several years and is very widely implemented. As we shall see, IPv4 has some serious shortcomings, and so a technically better alternative, IPv6, has been developed.

IP version 4

IP is a datagram routing protocol. An IP router connects to a number of subnetworks (which may be point-to-point data links, or multipoint subnetworks such as Ethernet LANs). IP addresses the problem of which subnetwork to deliver a datagram to, on the basis of the destination address that the datagram carries within itself. IP routing is insensitive to the class of service of the data in the datagrams that it handles; it has no concept of prioritization. Nor does it provide any error control. These issues are left for other protocols to address.

IPv4 addresses consist of four 8-bit numbers, for example 192.0.0.2. An IP address is divided into a part that identifies a subnetwork, and a part that identifies a machine within the subnetwork. Given the numbers of machines that want IP addresses, 2^{32} addresses leaves little to spare.

Because, on the one hand there are very large numbers of subnetworks to identify, and on the other hand some subnetworks contain very large numbers of machines, it would not do

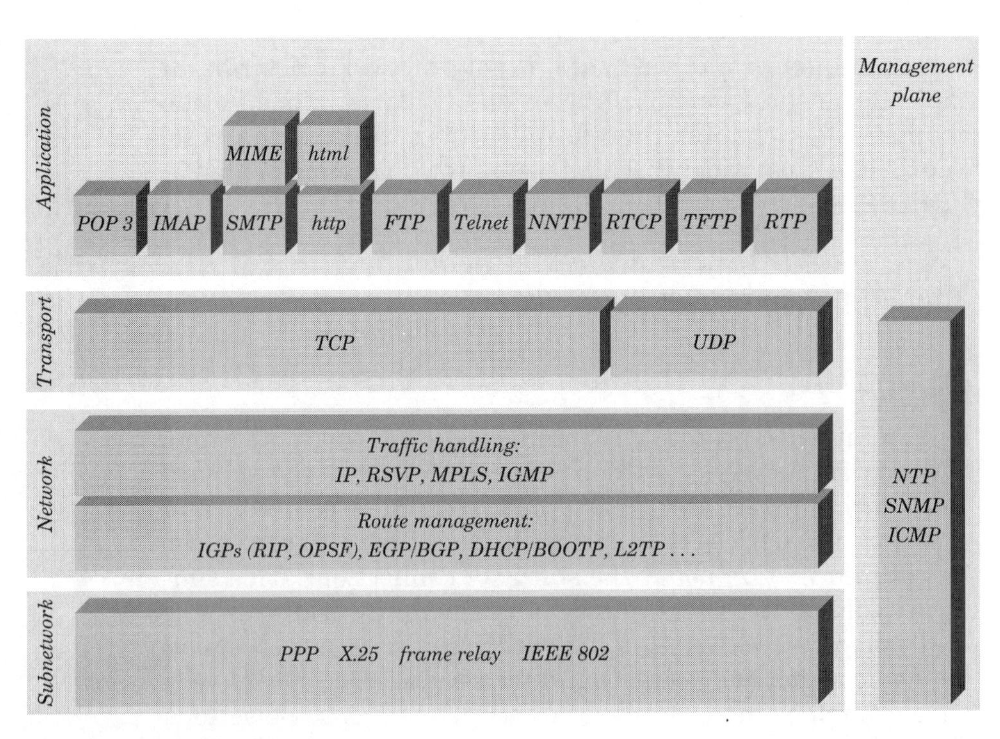

Figure 9.1 Selected internet protocols

to make a hard division of how big each of those parts is. For that reason there are four classes (A–D)[1] of IP address, each of which allocates different amounts of the address to locating the network and to locating the machine within the network.

There is also a newer and more flexible address format called *classless inter-domain routing* (CIDR, defined in RFC 1519, 'CIDR: an Address Assignment and Aggregation Strategy'). CIDR is not, however, supported by all internet route management protocols. It is supported by BGP-4 and OSPF, but not by the older EGP or RIP.

IP version 6

IPv6 (alias IPng)[2] addresses many of the limitations of IPv4. The cramped 32-bit address space is increased to 128 bits. The datagram header is made simpler, so that it can be processed in hardware and thus processed faster. The header also includes information about the QoS requirements of the carried data, so that prioritized handling can be applied.

However, at the time of writing, uptake of IPv6 is patchy, experimental and uncertain. This is because ways have been found to work around the limitations of IPv4. Network address translation gateway systems have mitigated the address space shortage. MPLS (*multiprotocol label switching*) and other prioritization systems have gone some way towards QoS management. We may have to wait until the pain of living with IPv4 becomes unbearable before we see widespread replacement of IPv4 equipment with IPv6 alternatives.

At the time of writing, uptake of IPv6 is patchy, experimental and uncertain.

Route management

IP routers need to know in which direction to route the IP datagrams that they receive. There are lots of internet protocols for managing the routing information.

For routing within an autonomous subnetwork, there is a family of *interior gateway protocols* (IGPs). For example, RIP, the *routing information protocol*, takes a fairly blunt approach.

[1] Well, actually five, as there is a Class E, reserved for future use.

[2] ng for *next generation*.

It has all the subnetwork's route information copied regularly to all the routers in the subnetwork (whether it is relevant to them or not), and then it gets the routers to do their routing on the basis of minimizing the hop count to the destination, regardless of conditions in the network. A more subtle alternative is the *open path shortest first* (OPSF) protocol, which only distributes information about changes to the routing information, and which supports an algorithm which takes into account the status of the available links, and which can guide the routing by assigning 'costs' to nodes.

Between autonomous subnetworks another race of routing protocols applies, including the *exterior gateway protocol* (EGP) family and the *border gateway protocols* (BGP, e.g. BGP4).

Address resolution

When an IP datagram reaches the destination subnetwork, the router there will need to know the physical address (for example, the Ethernet address) of the target host, so that it can route the final hop. The process of translation between IP addresses and *(sub)net point of attachment* (NPA, the physical address on subnetwork) is called address resolution. ARP, the *address resolution protocol*, works as follows. The router broadcasts the target IP address to all hosts on the subnetwork. The host that recognizes this as its own IP address replies with its IP address and NPA.

Most hosts can be expected to acquire their own IP address (through the *dynamic host configuration protocol*, DHCP, or the bootstrap protocol, BOOTP) and then to remember it. In some exceptional cases, however, a host is incapable of remembering its IP address, for example if it has no non-volatile storage. In such cases, the subnetwork has to have an address server which stores NPA/IP address pairs. Then the forgetful hosts can learn their IP addresses by using RARP, the *reverse address resolution protocol*. Here, the host sends a RARP message with its NPA to the address server, which replies with the right IP address.

Domain names

IP addresses are organized into domains. A domain is a collection of IP addresses, and is uniquely identified by a domain name. Domains are organized hierarchically. At the top of the hierarchy are the 'top-level domain names', each of which identifies a nation (.uk, .se, .de, .jp, .it etc.) or an area of common interest.

Second-level domain names identify IP subnetworks, for example pearsoned-ema.com, and can be translated into IP addresses through the domain name system (DNS). When a client process on some host needs to resolve a domain name, a name resolver process on the host uses TCP to pass a request to the *domain name server* in the same subnetwork. If this server does not have the answer, it passes the request on up a hierarchy of DN servers until the answer is found.

There are two different ways of doing this, known as recursive and iterative queries. Both methods traverse the DNS hierarchy. Iterative querying puts the processing burden on the resolver process, whereas recursive querying puts the load onto the name server.

Top-level non-national domain names

.com	commercial
.edu	educational
.gov	government
.mil	military
.net	Internet support

IP mobility

IP hosts are not always permanently connected to their subnetwork. A host with a dial-up connection could want to call into a number of subnetworks. For example, a large corporation will want its staff to dial into their nearest point of access to the corporate network, wherever they happen to be, to minimize PSTN charges. Or in mobile internet systems (such as UMTS), the user may be moving rapidly from one subnetwork to another.

However, there is a problem with this. An IP address refers to a specific subnetwork. If a device already has an IP address, how can it connect to another IP subnetwork? Datagrams that are sent to it will end up on its 'home' subnetwork, not the one that it has attached itself to.

As we shall see in the next few sections, a number of solutions to this problem are available. Also, we shall see how IP mobility raises issues of security, and thus introduce the related field of IP *virtual private networks* (VPNs).

Tunnelling

The usual solution to the problem of attaching an IP host to a foreign subnetwork is called *tunnelling*. The idea of tunnelling is to get the Internet to behave, for a selected stream of traffic, as if it were not a complex network with all sorts of routing possibilities but a point-to-point data link with only one possible route.

A tunnel is in effect a virtual circuit between two nodes, usually one on the roaming node's home network, and one on the network that it is visiting. IP datagrams that are addressed to the roaming node are accepted by the host on the home network, and passed to the host on the visited network *inside* other IP datagrams. The outer wrapper of IP datagrams (showing the IP addresses of the hosts at the ends of the tunnel) fools the Internet into shipping the roaming node's datagrams to the subnetwork to which it is attached. The following paragraphs describe two alternative approaches to tunnelling.

The layer 2 tunnelling protocol

L2TP, the *layer 2 tunnelling protocol*, is a recent development which aims to combine the best features of two earlier, proprietary systems: Microsoft's *point-to-point tunnelling protocol* (PPTP) and Cisco's *layer 2 forwarding* (L2F) protocol.

L2TP provides a tunnel for PPP (*see* Figure 9.2). The roaming node connects (say via dial-up modem or ISDN) to a *L2TP access concentrator* (LAC)[3] in the visited subnetwork. The LAC terminates the LCP sublayer of the PPP protocol, and does the dial-up authentication work. The LAC operates a tunnel through to a *L2TP network server* (LNS)[4] on the roamer's home subnetwork, hiding the IP addresses of the real client from the IP network. The LNS terminates the PPP NCP, and uses normal IP routing to connect the roaming node to wherever, just as if the roamer had a PPP connection direct to the LNS.

[3] Alias *network access server* in L2F.

[4] Alias *home gateway* in L2F.

Like L2F, L2TP offers no encryption of the tunnelled datagrams. If security is an issue, IPsec (discussed below) is a more appropriate alternative. And like PPTP, L2TP is not transparent; it requires the network's routers to support the protocol.

'Mobile IP' approach

The arrival of high-capacity mobile data communications technologies such as UMTS has given a further impetus to the development of IP mobility protocols. RFC 2002 (mobile IP) defines a mobile access system[5] which is different from L2TP, but which is also remarkably similar, at a coarse level.

Every subnetwork that accepts roaming nodes includes one or more *foreign agents*. When a roaming node attaches to a subnetwork, it uses extensions to ICMP router discovery (RFC 1256) to find out if it is in its home network, and if not, to find a foreign agent. Foreign agents regularly broadcast

Every subnetwork that accepts roaming nodes includes one or more foreign agents.

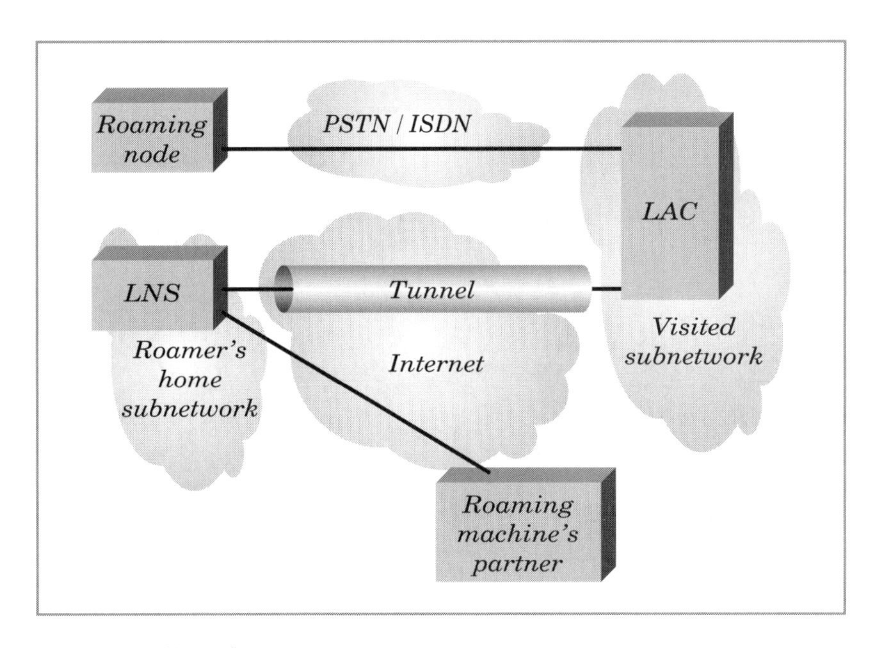

Figure 9.2 L2TP outline

[5] Mobile IP is discussed in more depth in Black (1999)

agent advertisements (as defined by RFC 1256), typically once every second. If that is not quick enough, the roaming node can also broadcast an enquiry.

A home agent in the roaming node's home subnetwork maintains a record of the roaming node's location, in terms of the care-of address of the foreign agent to which it is attached. A registration mechanism (over UDP) allows the roaming node to establish a session with a foreign agent, and to register its location with its home agent. This process is guarded by rigorous security procedures to prevent false registration. The home agent and the foreign agent tunnel messages between them, using IP-over-IP tunnelling (*see* Figure 9.3).

ICMP

ICMP, the *internet control message protocol*, provides a range (one might even say a rag-bag) of route management support services. ICMP provides routing error reporting, for example to report back to the sender that the destination is unreachable, or that the specified protocol is not present at the destination. ICMP supports reachability testing, where an ICMP 'echo' message is sent to, and returned from, a chosen IP address. The

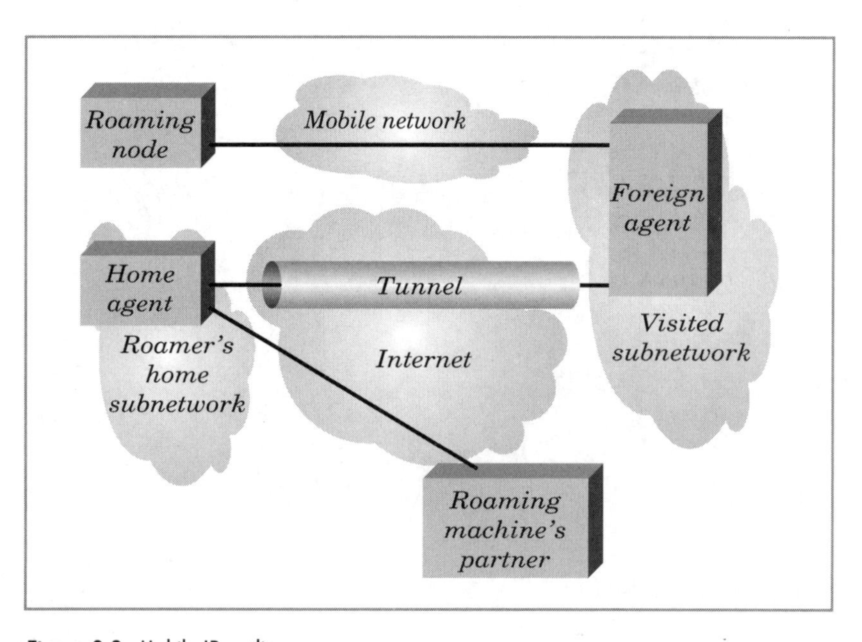

Figure 9.3 Mobile IP outline

ICMP echo facility is used by the Ping program, a simple network management tool, and also supports network transit delay measurement. ICMP supports network congestion control, including a source quench facility analogous to the call gapping discussed in Chapter 11. Also, ICMP provides route redirection facilities, for example to remove 'trombone' routes from the network.

Controlling IP performance

Because IP is a connectionless datagram protocol, there is no question of it being able to manage the QoS given to a particular stream of data. There is just no concept of a data stream, connection or circuit. Because IP datagrams do not carry information about the QoS that they require, IP routers cannot even apply prioritization at the datagram level. Still, as the internets are used for more and more demanding applications, network QoS becomes more important, and so there have been a number of attempts at addressing the problem of IP performance.

At one stage, it was expected that ATM would supersede IP, not least because of ATM's built-in QoS features. However, ATM is now seen as just another subnetwork technology, just another link in the chain. No matter how good ATM's QoS management is, it will apply only to the ATM subnetworks, and will not address the end-to-end problem. What is needed is a capacity management system at a level above the subnetworks, but which all subnetwork technologies can support.

RSVP

The *resource reservation protocol* (RSVP, defined in RFC 2205) enables a network user to reserve (and then release) capacity between two IP addresses. An RSVP-enabled network can then offer guaranteed QoS between those addresses, just as if it were a connection-oriented service, by expediting the progress of datagrams at each network node, according to their required QoS. However, to use RSVP, all the routers between the two IP addresses need to be RSVP-capable. RSVP has yet to be widely implemented on public IP networks, and is usually restricted to private internets.

MPLS

Multiprotocol label switching, alias *label switching*, alias *tag switching*, alias *level 4 switching* works by getting the applications that are using the network service to ask the edge devices for a particular quality of service. The edge devices then insert a label in the head of each frame, to indicate the QoS required. The implementation of the labels depends on the subnetwork technology. So, for example, in an ATM subnetwork, the label replaces the *path* and *circuit* numbers in the cell header. Then it is possible for all the devices in the network to give differential service depending on the label value, without having to remember complicated associations of IP address pairs and QoS. However, this requires that all the devices in the 'MPLS-enabled' network should be modified to implement the label switching system.

MPLS originated as a proprietary protocol from Cisco, and has been presented to the IETF for issue as an RFC.

Diffserv

The IETF is also looking at an alternative called *diffserv*. Diffserv offers differentiated service performance, rather than a guaranteed QoS. Diffserv works by modifying the layer 3 datagram header (which already includes, in IPv4, a differentiated services field). As diffserv does not interfere with the layer 2 frame structure, it requires less modification to the subnetwork technology, but still requires that all the layer 3 routers should support it.

Whether IP QoS management will in the end be done through diffserv, MPLS or IPv6 is, at the time of writing, anybody's guess.

Clock distribution

Clock distribution in IP networks is effected through NTP, the *network time protocol*, which works hard to ensure accuracy of distribution despite the variable-delay transmission networks that it has to operate over.

IGMP

The *internet group management protocol* (IGMP) provides network level services for group multicast. Users can send a notification that a multicast is starting, can send the multicast data stream, and can ask to join or quit the multicast. A number of IP network operators have developed a separate backbone network for multicast distribution, called the *multicast backbone* or *Mbone*. The Mbone offers 500 kbps capacity, which is used for example for distribution of live concerts. The Mbone uses specialized M-routers, which copy each datagram they receive to every downstream branch of a defined routing tree.

IP network operators have developed a separate backbone network for multicast distribution, called the Mbone.

Internet transport protocols

At the transport layer, there are just two internet protocols worth knowing about: TCP and UDP.

TCP

The internet *transmission control protocol* (TCP) offers a connection-oriented 'reliable stream transport service' similar to an OSI class 4 transport service. TCP provides error detection, retransmission of lost or errored data, notification of delivery and resequencing of jumbled packets on arrival. TCP also offers an 'urgent' option, like the expedited delivery service offered by the OSI transport service.

A TCP end point is an IP address plus a 'port' number. The port number has nothing to do with physical ports. It is just a way of identifying the destination application within the target machine. Several port numbers have been allocated to standard application layer protocols, under the banner of 'well known port numbers'. For example, port number 21 is for FTP, 22 is for Telnet, and 25 is for SMTP.

UDP

The *user datagram protocol* (UDP) offers a connectionless service. It will convey a single PDU between two end points. Therefore the full UDP addresses (which again comprise IP address and port number) of both ends have to appear in each datagram. UDP includes end-to-end error detection, but offers no assurance of delivery. It is essential to the understanding of applications that use UDP, that UDP is essentially unreliable, and a message sent via UDP may just vanish without trace if, for example, it passes through a network node that is overloaded and having to dump excess traffic.

Internet data application protocols

File transfer

The internet *file transfer protocol* (FTP) provides remote file manipulation services over a TCP transport.[6] Files can be pushed (sent) as well as pulled (downloaded), and multiple concurrent file transfers are possible. FTP provides facilities for working with three kinds of file:

- a structured file is a sequence of records of fixed length, composed of a small range of standard types (binary, ASCII, and EBCDIC);
- a random access file is similarly structured, but includes variable-length records;
- an unstructured file is a binary bit stream, which is transferred transparently by FTP.

The *trivial file transfer protocol* (TFTP) is a simpler protocol that is sometimes used as an alternative to FTP, generally across LANs. For example, TFTP is often used for loading operating system images onto thin client devices. The low error rate expected from LANs enables TFTP to use UDP rather than TCP. Error checking is done at the application level (as part of the TFTP protocol), but because of the short transit delay

[6] Where multiple file transfers are proceeding in parallel, there is usually one TCP connection for each file, and one more for supervision.

across a LAN, a very simple error checking protocol, with a window of one, is possible. Since TFTP has no authentication mechanism, it is usually restricted to a bounded LAN.

Email

The *simple mail transfer protocol* (SMTP) supports the transfer of electronic mail between mail servers. SMTP uses TCP, and defines the well-known email address format JohnSmith@ieee.org. A related RFP defines the header keywords which email programs are required to supply (e.g. To, From, Reply To, CC, Subject, Date and Encrypted) plus the extras that are accumulated along the way through the network (e.g. Received from). SMTP dumps the emails into mailboxes on mail servers. The users then must use either POP3 (*post office protocol* 3) to download their mail from the server to their own machine, or else IMAP4 (*internet message access protocol*) to browse mail on the mail server. Both POP3 and IMAP4 use TCP to access the mail.

SMTP is based on transferring messages as ASCII characters. Where non-ASCII files need to be transferred with an email, extended formats defined in the *multipurpose internet mail extensions* (MIME) are used.

The World Wide Web

The World Wide Web (WWW) has been the 'killer application' which has made data communications interesting and accessible to the masses, and which has been a major driver of the current shift to IP networks.

A Web page is addressed by its *uniform resource locator* (URL), which takes the form <protocol>://<domain name>/<filename>/ as in http://www.iee.org/flashpoint.html/

Web pages are defined in a number of markup languages (such as HTML, the *hypertext markup language*), and accessed via corresponding application protocols (most commonly http, the *hypertext transfer protocol*) over TCP.

HTML was the first markup language to be widely implemented over the Internet, and the great majority of Web pages and browsers use a combination of HTML, embedded object types such as MPEG, JPEG or MP3, and portable applets written in Java. However, HTML is limited in a number of ways, and there are other markup languages which offer to improve on it. Generally, all these markup languages are defined using the same metalanguage, SGML (*standard generalized markup language*), so they all look quite similar.

HTML is restricted to defining a Web page from the point of view of a human user. So while HTML can say things about how an item should appear, it can say nothing about what the item means. The assumption is that the human user can work that out from the context. XML, the *extensible markup language*, introduces a layer of syntactical information, so that the browser can understand more about the data and present it in a way which suits both the data type and the browser's particular environment. XML can also be used to define the syntax of data structures for exchange between applications without human interpretation.

HTML is restricted to describing page layout in two dimensions. VRML, the *virtual reality markup language*, can define three-dimensional spaces and objects, which a browser can present either in three dimensions through special virtual reality equipment, or as a two-dimensional perspective view on a screen.

HTML is oriented to visual display of information, which limits its applicability to situations where the user can spare the attention to look at a screen. VoxML™ is a language based on XML, which is oriented to presenting information through speech synthesis and speech recognition. Either the user can have his or her own VoxML browser embedded in a Web access device, or they can use normal voice telephony services to access a VoxML browser elsewhere.

A trivial VoxML™ page

```
<?xml version = '1.0'?>
<DIALOG>
<STEP NAME> = 'salutation'>
<PROMPT> Hello, world
</PROMPT>
```

Telnet

Telnet is a TCP application which provides remote bi-directional character mode terminal access. Telnet is based on ASCII character codes. If the terminal is not an ASCII terminal, its Telnet client has to convert its character code to ASCII for transmission. Similarly, if the host does not talk ASCII, the host-end Telnet server (sometimes called the pseudo terminal) has to do the same.

RTP

The *real-time protocol* (RTP, defined in RFC 1889) offers a time-transparent connective service, usually on top of UDP. RTP sessions are managed via the *real-time control protocol* (RTCP, also defined in RFC 1889).

SNMP

SNMP, the *simple network management protocol*, is discussed in Chapter 11.

Internet telephony

Once upon a time, there were data networks, which were packet mode, and there were voice networks, which were circuit mode. Now, a lot of voice traffic is carried over packet mode data networks. Sometimes this is called *internet telephony*, sometimes *IP telephony*, sometimes *voice over the net* (VON), and sometimes *voice over* IP (VoIP).

Why would people want to do such a thing? It is not a question of doing it to get digital-quality voice transmission, because circuit mode networks are already digital inside. And it is not as if it is easy; it is very difficult to get 'toll-quality' voice transmission over packet mode networks.

For network users, Internet telephony offers long-distance calls for the cost of the local call to their ISP.

For network users, Internet telephony offers long-distance calls for the cost of the local call to their ISP. And in some territories, local calls are free. For network operators, internet telephony offers a way to evade licence regulation. In many territories, the provision of circuit mode voice services is closely controlled, but packet mode voice services fall outside the regulators' net.

For network operators, internet telephony is also, in the long term, a less wasteful use of capacity. In a circuit mode network, each call consumes 56 kbps or 64 kbps of network capacity, regardless of whether there is any sound to convey. Internet telephony tends to use more efficient codecs which require less capacity, and which do not use up network capacity during the gaps in the conversation. Internet telephony thus collapses voice services more efficiently onto the underlying transmission network.

Internet telephony elements

There are several competing sets of standards for internet telephony, from the ITU-T and from the IETF. Figure 9.4 shows the ITU-T model, which is the oldest and currently the most widely deployed. The earliest internet telephony systems were based on personal computers. For Internet telephony, a PC needs to be equipped with data connection to an ISP (typically over a modem), and an audio card connected to a telephone. The PC

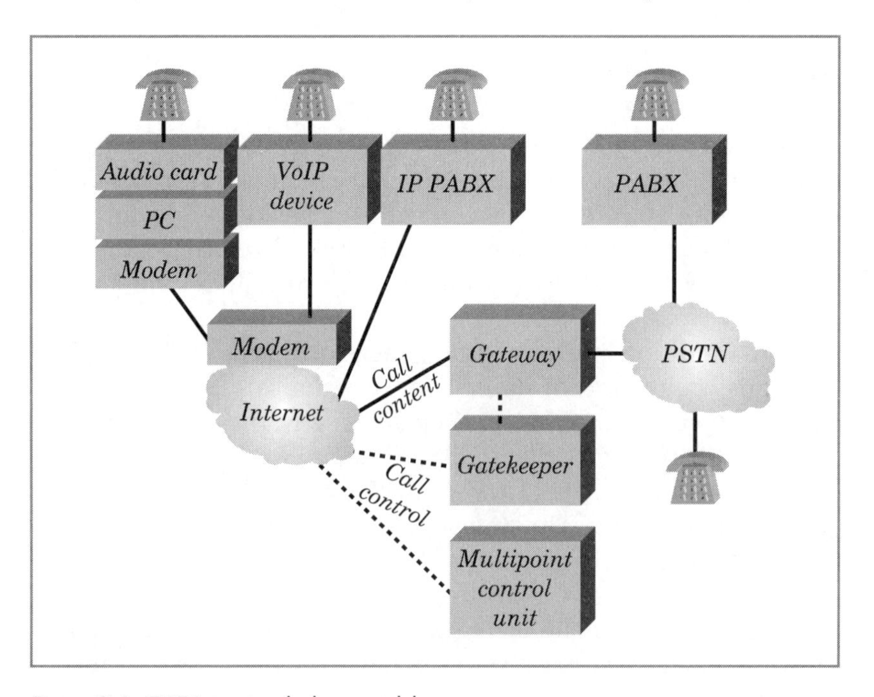

Figure 9.4 ITU-T Internet telephony model

software needs to include an audio codec and a VoIP protocol. Alternatively, the codec can be implemented in hardware on the audio card. Such an arrangement is sometimes called a 'soft IP phone'.

More recently, 'hard IP phones' have been developed. These have a connector for a telephone, and a connector for a telephone line. They contain an audio codec, an IP telephony stack, and a modem. Also, several PABXs have been equipped with Internet telephony interfaces in addition to their normal circuit mode connections.

In all these cases, the equipment makes an IP connection to another Internet telephony device, establishes a session, then transfers the call content as packetized codec data.

However, as the Internet telephony community is a minority of telephone users, an Internet telephony solution that could only support traffic among that community would be of very limited interest. To bring Internet telephony to a wide market, it needs to be able to interwork with the ordinary PSTN. To enable PSTN interworking, two additional devices are used: a *gateway*, which handles the call content, and a *gatekeeper*, which manages aspects of call control. For conference calls, a third device, a *multipoint control unit* (MCU), is also required. These devices' functions are discussed below.

ITU-T VoIP standards

The umbrella ITU-T recommendation that governs internet telephony is H.323 rev 2 (1998), 'Packet-based multimedia communications systems'. This defines mechanisms for carrying not only voice but also data and video streams over IP networks (*see* Figure 9.5).

H.323 calls up RTP/RTCP[7] as the content bearer protocol for audio and video streams, and H.225.0 for call signalling protocols and media stream packetization. H.225.0 includes the *registration, admission and status* (RAS) protocol, which operates between a call end point and a gatekeeper, for gatekeeper discovery, and to register the end point with the gateway. It also includes a *call signalling protocol*, which is a subset of Q.931, carrying call control messages between the end points and gatekeepers.

H.323 Internet telephony calls up Q.931 for call control (so

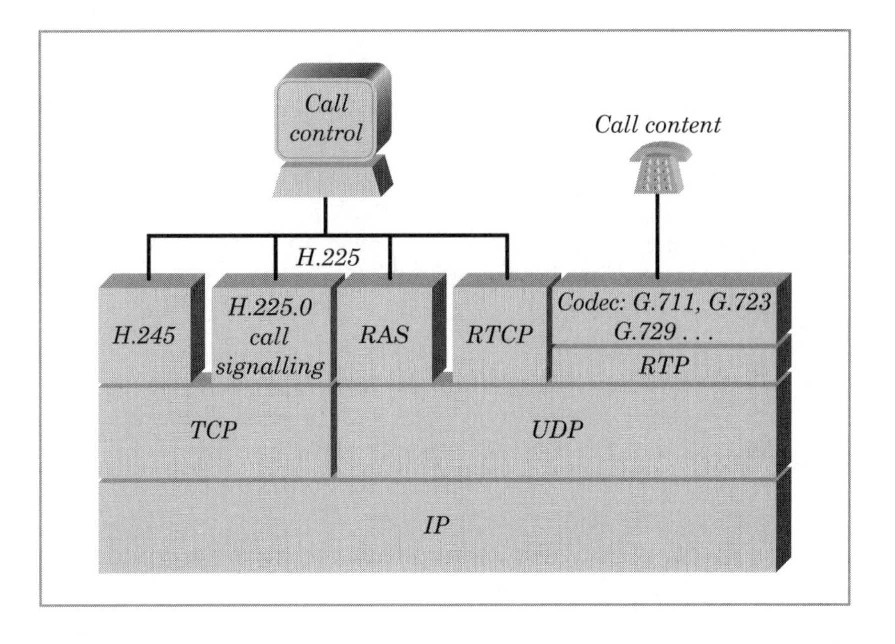

Figure 9.5 H.323 protocols

when an IP phone is calling an ISDN phone, the IP phone can use the ISDN advanced signalling facilities). It uses the H.245 control protocol for multimedia communication, to negotiate media requirements (such as audio codec selection) and open bearer streams (over UDP).

ITU-T recommendation H.324 describes a gateway for connection to a PSTN, while H.320 does the same for an ISDN.

PSTN/ISDN side	function	internet side
G.711 56/64 kbps codec	transcoding	G.723 codec[8]
T1 or E1	transport	RTP, UDP, IP
SS7, Q.931	call control	H.245, H.225, Q.931
E.164 telephone numbers	addressing	IP addresses

[7] In real life, however, RTCP is not always implemented.

[8] H.323 specifies G.729, but G.723.1 has become established in practice because it requires less channel capacity.

IETF SIP model

Recently the IETF has come up with an alternative set of standards for Internet telephony. In the IETF model, H.323 is replaced with the *session initiation protocol* (SIP, defined in RFC 2543) at the application level. SIP supports call control and address translation to IP addresses from SIP addresses. It runs on a combination of UDP and TCP, and also uses the services of RSVP, RTCP and RTP.

SIP addresses are of the forms

username@domain,
or

phonenumber@gateway,

for example
+44 1225 444 888@bt.com;user=phone.

Comparison of ITU-T and IETF models

H.323 reached the market three or four years ahead of SIP, and also offers richer facilities for selecting and negotiating codec options. It is not clear, however, that these advantages have given the ITU-T model enough market momentum to ensure its dominance.

SIP offers a number of technical advantages. It is radically simpler; far fewer standards are involved, and there is much less scope within the standards for incompatible alternative interpretations. SIP is inherently more extensible. It is designed so that SIP devices can add protocol elements, and receiving devices can ignore the ones they do not recognize. SIP offers better support for personal mobility. For example, a SIP call attempt can try to contact the called user at several locations simultaneously. SIP holds all the information about the state of a call in the devices at the network edge, in contrast to the concentration of call information in the ITU-T gatekeeper. SIP is therefore more scalable.

SIP offers a number of technical advantages. It is radically simpler

VoIP carrier network standards

Within VoIP carriers, alternative sets of standards are used to address the issues of scaling and network control. Three sets of standards are currently in contention:

- *media gateway control protocol* (MGCP) from the IETF; this has a combination of features from two earlier standards sets: *internet protocol device control* (IPDC) and Bellcore's *simple gateway control protocol* (SGCP).
- *network-based call signalling* (NCS), an MGCP derivative aligned with the DOCSIS cable modem standards;
- ITU-T H.248 (alternatively called *Megaco* by the IETF), another MGCP derivative.

Internet telephony network performance issues

Where internet telephony is tried over the public Internet, it is liable to severe QoS problems. Call setup can take 30–45 seconds. The network transit delay is often around 0.5 seconds, and is subject to severe timing jitter. Because of the impossibility of securing capacity across the Internet, packet loss is common. These difficulties mean that reliable toll-quality telephony across the Internet is not possible.

Internet telephony has been far more successful, so far, within corporate intranets, where the network can be dimensioned and traffic can be managed, so that there is considerable over-capacity in the network and a better and more reliable service can be provided. A number of telcos have taken the same route, offering commercial Internet telephony services over their own closely managed networks.

Internet facsimile

An ordinary fax machine scans the page, digitizes the image, and then transmits it over an integral 14.4 kbps modem across the PSTN. Within the PSTN, the modem warbles are themselves digitized, and carried over a 56 kbps or 64 kbps channel. Clearly, there is enormous waste of capacity, which is addressed by *facsimile over IP* (FoIP) technology. The PSTN and Internet alternatives are shown in Figure 9.6.

The ITU-T recommendations for FoIP are T.37 and T.38.

They are based on the ordinary T.30 fax standard, plus RFC 2305 (IP fax protocol) and RFC 2301 (file formats). T.37 provides a basic real-time service. T.38 offers additional store-and-forward and multishot facilities.

FoIP is implemented through a FoIP *interworking function* (IWF), which reverses the fax modem to retrieve the digital stream, and provides the Internet connectivity.

Packet mode telco network architectures

The introduction of packet mode technology into telco networks has enabled a number of simplifications and improvements to the telco network architectures.

Transmission network reduction

One improvement is the possibility of using packet networks to save on backhaul. Backhaul is a loosely used term for the part of the transmission network that is just there to get the traffic from the user to a switching node. Sometimes that just means

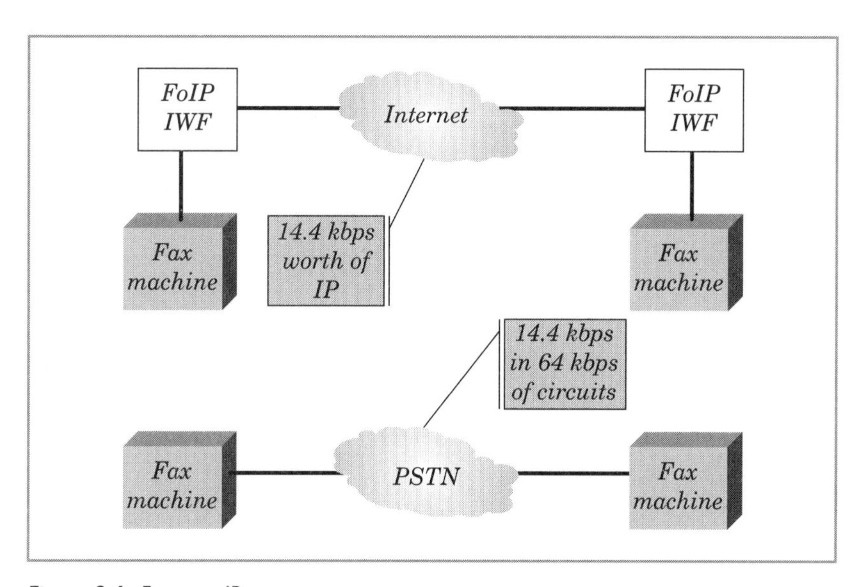

Figure 9.6 Fax over IP

the local loop, but not always. Figure 9.7 shows three possible voice network architectures.

Let us imagine a telco that has to serve hundreds of thousands of customers, who are distributed over a large territory. The customers all want network access, but most of them only actually use the network sparsely. The whole call traffic for the territory could be handled by one large telephony switch, and by buying a big one, the telco could save both capital and operational costs, compared with having lots of little ones. Option (a) in the figure shows this approach. The trouble with this is that all the customers' connections have to be backhauled to the switch, and even if the switch were placed centrally in the territory, the transmission costs would be outrageous. Also, the transmission network would be grossly under-used, as the line from a customer to the switch would be used only when the customer was making a call.

So real-life networks up to about 1998 had to take option (b), of having a lot of small switches placed close to the customers, partly to handle local traffic, but mostly to reduce the size of the backhaul network. The local switches multiplex the traffic onto the transmission network, so that it is only carrying calls.

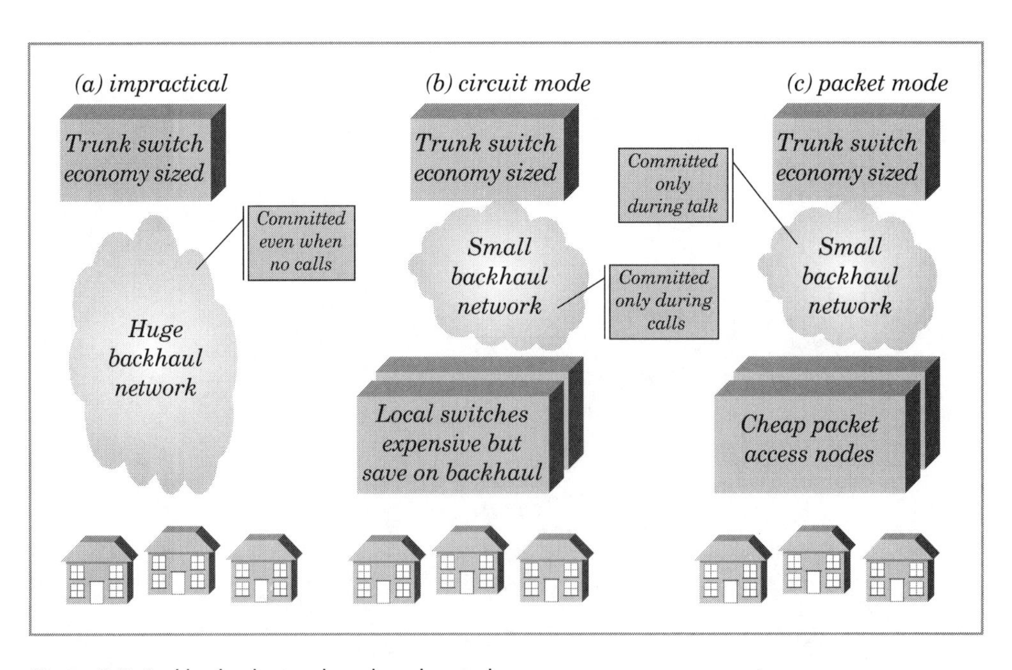

Figure 9.7 Backhaul reduction through packet mode

The introduction of packet mode transmission networks (originally ATM, but now other IP network types) offers a third approach, (c). The customers are connected to cheap packet network access nodes, which handle the local loop signalling but offer no switching. These access nodes use Internet telephony protocols to pass the traffic back to the one large central switch. The backhaul network can be even smaller than in case (b), because it carries packets only when there is actually some sound to be conveyed. Where the customers include teenagers, the savings from not transmitting long meaningful silences may be significant.

Inverse multiplexing

Another neat trick which packet mode networks enable is inverse multiplexing, illustrated in Figure 9.8. Suppose a customer wants a high-capacity link between two points, and the telco, although it hasn't got a spare high-capacity link, can offer a number of lower-capacity ones. With circuit mode multiplexing techniques, this is no good; there is no way of sticking together several lower-rate channels to emulate one higher-rate one. But with a packet mode network, the telco can put a packet switch at each end of the connection. These packet switches can offer the customer a single high-rate connection, and multiplex the traffic from it over several parallel links, recombining the packets in the correct sequence at the far end.

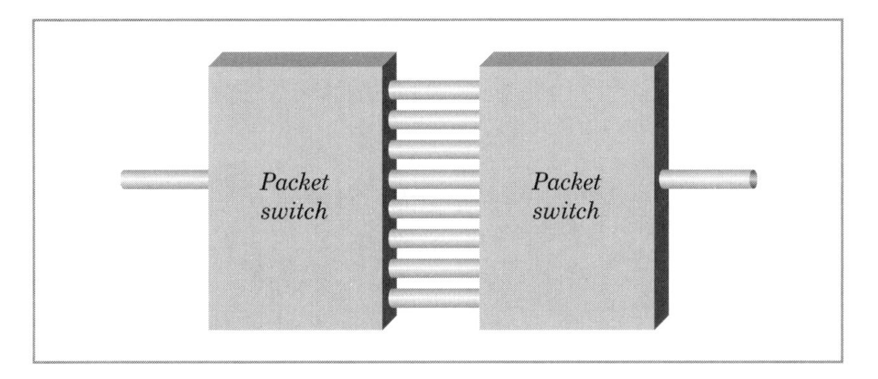

Figure 9.8 Inverse multiplexing

Transmission network architectures post-IP

The availability of IP and other packet mode technologies has introduced many new possibilities for transmission network architectures, as Figure 9.9 shows. Data teleservices are moving rapidly away from circuit mode networks (leased lines), to IP networks. Voice teleservices run largely over circuit mode networks, but a small and increasing fraction of them run over IP networks.

Circuit mode telco networks still include significant elements of PDH subnetwork technology, but the majority of circuit mode multiplexing is now SDH or Sonet. Where ATM is used, it tends to be as a flexible overlay to an SDH network. SDH and PDH both run over optical fibre and also over WDM systems (and so, less commonly, can ATM).

IP telco networks use either ATM subnetworks, or PPP. The PPP can be carried over SDH/Sonet or else it can run 'directly' over WDM. This PPP over WDM architecture is called *IP over glass*, or *IP over lambda*. Really, it is something of a fraud; the IP does not run directly over the glass, it runs over PPP. And the PPP does not run directly over the glass either; it uses one

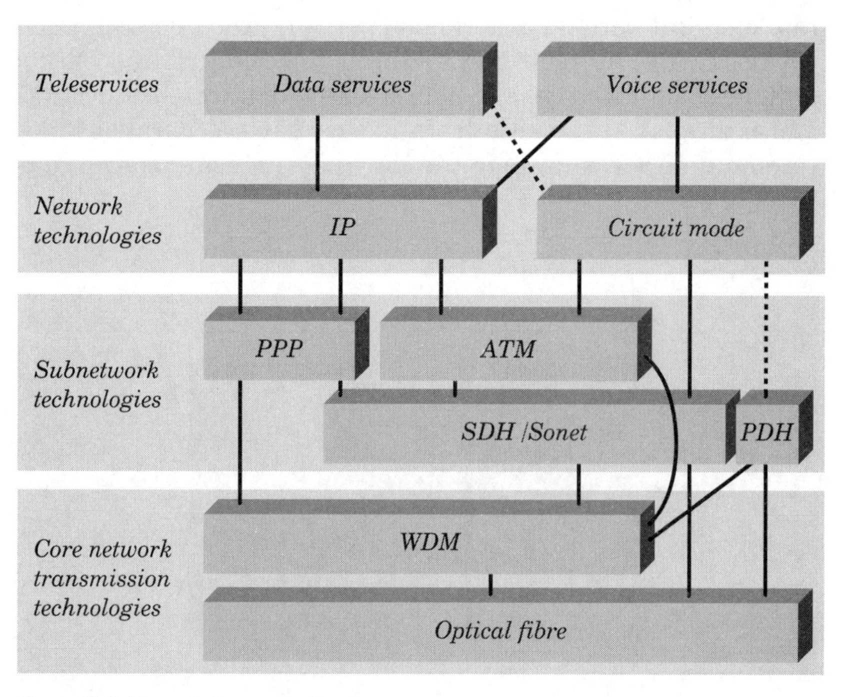

Figure 9.9 Transmission network options

of a number of proprietary protocols, which are stripped-down versions of Sonet or SDH, providing the necessary frame structure.

Some Internet network architectures

Figure 9.10 illustrates some of the many network architectures that are evolving as telco networks transform themselves to support, or use, IP technology. One domestic customer is shown connected to his or her ISP via a telco's circuit mode voice PSTN, with a modem at the customer's end and a modem bank at the ISP. Another domestic customer has a DSL local loop, terminating at a DSLAM in the local telco's local exchange building, where their ISP has its POP located. Several of the ISPs, and a business customer, are connected via an ordinary telco's IP network offering, which offers various interfaces including ATM and frame relay, built on top of a pre-existing SDH network. Another telco offers IP services only, and runs IP over lambda.

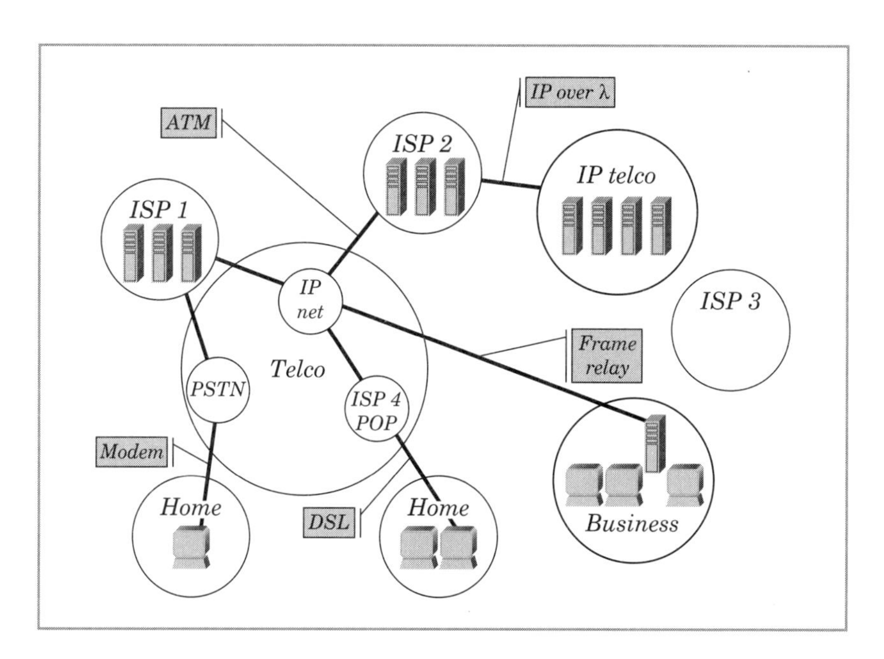

Figure 9.10 Telco and Internet architectures

The Internet access bottleneck

The Internet has brought trouble for telco network planners. Telco networks were designed, until the mid-1990s, on the assumption that almost all the dial-up traffic would be voice traffic, and that it would have call-holding times of a few minutes, and a certain mix of local and trunk calls. Now Internet access is skewing the traffic pattern. Internet access produces exclusively local calls, with call holding times measured in hours. Line demand has increased, and so more transmission and switch fabric is required, yet the local nature of the traffic means that this investment brings little or no increase in revenues for the telcos. The congestion in the overloaded telco networks disappoints not only the callers but also the ISPs, which need their customers to be able to access them.

The Internet has brought trouble for telco network planners.

Planning aside, there is an intrinsic wastefulness in using full-duplex circuit mode network resources for bursty and asymmetrical packet-mode traffic. The longer the circuit mode link between the Internet user and their ISP (the point at which their traffic is transferred into a packet mode network), the worse the waste is. Therefore, the solutions to the Internet access bottleneck tend to push the ISP POP towards the customer, for example co-located with RCUs.

Further reading

Again, Halsall (1995) provides a good overview. For more detailed treatment of IP protocols, Black (1999) is helpful. For VoIP in particular, Douskalis (2000) is excellent. Electronic commerce, while really outside this book's scope, is covered authoritatively and easily by Norris *et al.* (2000).

10 Telco business processes

A telecoms network operator is a business like any other, and therefore many aspects of its business are unremarkable. This chapter sets aside those many aspects, such as customer relationship management and corporate finance, which are common to most businesses, and focuses on the areas where a telco is significantly different from the rest. These areas are also the ones in which a telecoms software engineer is most likely to be interested.

Modelling the business

All telcos are different, but most are similar enough to enable a high-level generic model of their typical business processes to be built.

The best such model, so far as depth of detail and widespread usage go, is the *telecom operations map* (TOM, TeleManagement Forum (1999)(a)). TMF, in an initiative called SMART, sent emissaries round a whole lot of telcos to ask them about their business processes. They put the results together into the TOM, which provides a well-known process model which can be applied, more or less, to most telco operations. The TOM is supported by a second level of detail, in the *network management detailed operations map* (TeleManagement Forum (1999)(b)). However, while these documents are useful when doing telco business process engineering, even the TOM is too detailed for our purposes here. I want to present a model of telco business processes that is simple enough for you to carry around in your head: the eight-box model in Figure 10.1.

A *service product* is defined. In parallel with the selling process, the network platform for the service is *designed* and *built*, and goes into *operation*. When a sale is made, the service product is *provisioned* for the customer. Then the customer's service usage is *metered*, and the quality and integrity of the service has to be assured. The metering information feeds the process of *billing* the customer and *collecting* the revenue, which is generally what the telco is there to do. Lastly, the billing process may be used to lead to further sales, as we shall see. Of course, there are usually multiple service products and many customers, so the model shown actually operates in many concurrent instances. The following sections introduce what goes on in each of the functions shown.

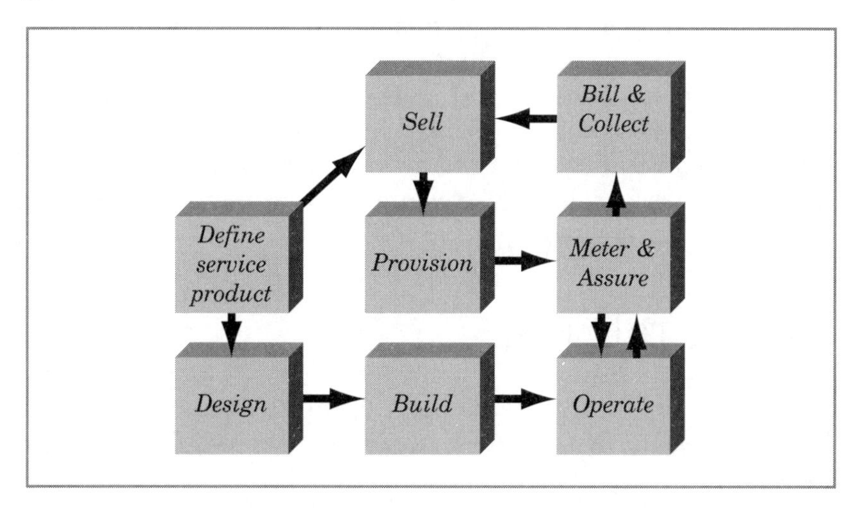

Figure 10.1 Telco business processes

Defining the service product

Defining a service product includes several elements, of which the network technology is only one. A service product is defined by marketing people, as well as engineers, in terms, at least, of:

- the network service: what the technical network offering is. An example might be the ability to associate two numbers with a single customer line, and deliver distinctive ringing cadences according to which number is called;

- the presentation of the service: whether it is to be offered as a single item or as part of a bundle; what it is to be called; how it will be drawn to customers' attention. In this example, we could offer it in a bundle with a discount rate for high line usage, advertise it on mid-evening TV, and call it 'Your Call';
- the tariffs for the service: whether it is chargeable by usage, rental or one-off fee. In this case we might judge that it is a novelty item that will generate secondary revenues through increased line usage, and charge it on a one-off basis;
- the bill and collect options: will it be prepaid or credit-based? In this example we could decide to add the charge to the quarterly phone bill;
- most importantly (and in real life, not last in the list), the target customers, in terms of geographical or demographic group. In our example, the targets are existing subscribers aged 34–55, the group most likely to have teenage offspring receiving tiresome numbers of calls.

Designing the network platform

Designing the network platform for a service product clearly depends very much on what kind of network there already is (if any), and what kind of service product is required. However, there are some common steps which most, if not all, network planning tasks go through.

Forecast the connections and traffic

This is as much a marketing job as an engineering one. The planner will consider the demographics of the target user group, and estimate the anticipated service uptake, over time and across the regions where the service is to be offered.

Locate the network nodes

Whatever the service, it will depend on one or more kind of network nodes. These may be switches, IN boxes, call centres, billing systems, or other things. The planners derive a design for the number and location of the nodes and the boundaries of the areas that each node serves, locations and boundaries, from the forecast traffic and from the constraints of the technology.

Sometimes the technology constraints are easy to model, and the only uncertainty is in the service uptake estimates. However, sometimes the technology is itself difficult to estimate for. An example of this is wireless network technology.

Wireless coverage can be planned by statistical extrapolation from previous experience of similar terrain, or by attempting to model the radio propagation qualities of the proposed cells. For example, one can attempt to model the radio properties (reflection, dispersion and refraction) of all the obstacles in the area, such as hills and buildings. Then, by using the same kind of ray-tracing technique used for producing lighting effects in 3-D computer models, the likely effect of the known obstacles can be estimated for every part of the area. While this approach has much appeal, it is limited by the accuracy of the planner's detailed knowledge of what obstacles there are, and the accuracy with which their radio properties are estimated. A new building, or a building of an unexpected type, can introduce significant errors into the estimates.

However coverage is planned, it is always subject to error, and cellular network designers expect to have to iterate towards perfect coverage, on the basis of live trials once the network is built. While gaps in coverage are irritating for mobile wireless users, they can be catastrophic for fixed wireless users. A serious problem for fixed wireless network operators is that it is difficult to be sure, in advance, whether the network coverage will be satisfactory at any particular house. This in turn makes the task of selling fixed wireless services peculiarly difficult.

However coverage is planned, it is always subject to error.

Determine the traffic distribution

From the chosen node locations and forecast traffic, the planner can have a go at estimating the traffic distribution between the nodes. Again, this can be straightforward where the network service is well understood, but can also bring nasty surprises. For example, one large telco decided to introduce a voice mail service. To limit its risk in introducing the service, it invested in only a few voice mail systems, dotted around its territory. When it launched the service, it found to its horror that whereas previously a call to an engaged user had taken up very little network capacity, now it incurred:

- a call leg to the called user's exchange;
- a call leg from there to the nearest voice mail box;
- later, a call from the called user to retrieve their voice mail;
- often then a call back to the person who left the message.

It is easy for us to see this now, but at the time the technology was new, and its implications, particularly for call attempt rate, just were not thought through well enough. The telco concerned found the new service chewed up network capacity so much that it had to withdraw it from sale until it could install a lot more voice mail boxes.

Determine the traffic routing

Once the planner has an idea of how much each node will want to communicate with each other, he or she can devise a routing plan, showing what path through the network each node pair will be connected over. The planner can then calculate (as discussed below) appropriate capacities for each inter-node circuit.

Dimension and cost the equipment

Now that the planner has an idea of how much traffic capacity is required, he or she can, in principle, finish off the job by dimensioning and costing the nodes and transmission plant, and doing detailed plans for their physical installation and support. In real life the process is much more iterative, because there will be constraints on the equipment capacity (it only comes in certain sizes), financial factors to consider (such as a special deal offered by a manufacturer of marginally unsuitable equipment), and other external constraints (such as the planning difficulties of putting a switch node in the middle of Piccadilly Circus, where the network traffic naturally focuses).

The challenges of network planning

Network planning is very difficult. The equipment and work planned usually have very high cost, and need to achieve a very long installed lifetime, say 20 years, to recoup their cost. Service uptake of a new service is always uncertain, particularly where there is free competition. And network operators are having to change the services offered rapidly, so equipment installed for service A is likely to have to be used for the foreseen services B and C, and probably the utterly unimagined services X, Y and Z.

Teletraffic design engineering

There is a whole science of dimensioning network equipment to match anticipated load, generally called teletraffic engineering. Teletraffic engineering evolved over the 80 years or so when circuit-switched traffic was the only sort. It is, therefore, so far as circuit mode networks go, a mature and well-understood business.

The basic goal of teletraffic engineering is simply to reduce the probability of call blocking to some chosen level. Meeting that goal is a complex job, for a number of reasons:

- call attempt rates are not generally deterministic. While it may be statistically true that the residents on my little street will generate 20 call attempts between the hours of 12:00 and 13:00 on an average Tuesday, on a particular Tuesday they may all independently decide to make their calls in the five minutes between 12:30 and 12:35;
- while one can estimate future traffic by extrapolation from previous periods, atypical events may cause huge departures from normal calling patterns. For example, television phone-ins may cause as many as 12 million call attempts in one hour; unpredictable disasters lead to floods of calls from anxious friends;
- where traffic passes across a number of operators' networks, changes in operator A's network routing can cause unexpected changes in the traffic on operator B's network;

- the practice of multiplexing traffic onto a small number of high-capacity trunks and nodes exacerbates the problem, firstly because (as we shall see below) a high-capacity trunk responds less gracefully to overload than a low-capacity one, and secondly because a single failure of a high-capacity network component can take out a relatively large part of the network's capacity;
- as we saw in Chapter 7, some network designs respond to overload by becoming less and less efficient, risking instability and total network failure;
- a network plan also needs to offer some level of resilience to operational problems. For example, a mistake in a switch's datafill may cause misroutes and resulting extra circuit legs, or even circular routes.

A network plan needs to offer some level of resilience to operational problems.

Erlangs and call attempts

Circuit switched traffic is measured in two ways: *erlangs* and *call attempts*.

The erlang[1] (E) is a dimensionless unit, which describes the volume of traffic across a bearer network. The traffic on a trunk, in erlangs, is defined as the average call holding time multiplied by the average call setup rate. Figure 10.2 shows a trunk carrying five transmission circuits, but only two erlang of traffic. Mr Erlang, who was keenly interested in the heavy

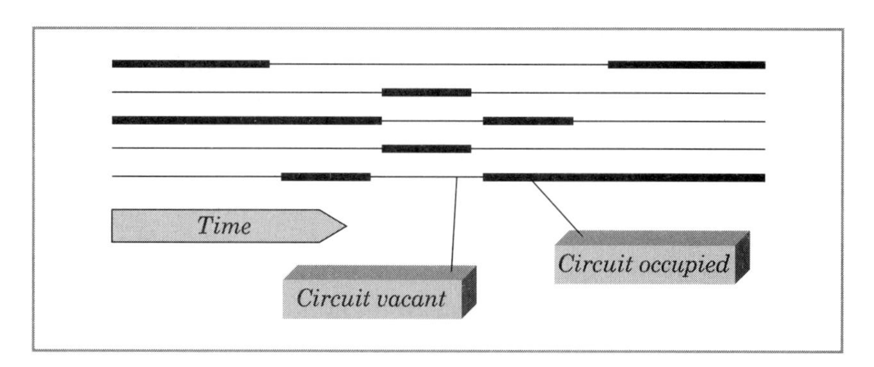

Figure 10.2 Two erlangs of traffic on a five-circuit trunk

[1] Named after Agner Krarup Erlang, the Danish mathematician.

maths of queuing theory and Markov processes, gave the world a number of handy teletraffic design tools. For example, Erlang's B loss formula gives the probability of call blocking for a given number of circuits and a given offered load.

The other way to measure circuit mode traffic is by call attempt rate. Call attempts use the signalling network and the call control systems (both on-switch and IN), which can be dimensioned separately from the bearer network. Also, call attempt rates are different from call setup rates, and the two are becoming more and more independent as IN services develop.

Network dimensioning

Networks are typically dimensioned on the basis of the required blocking performance for the *peak busy hour* traffic level. This is derived by looking at the daily traffic distribution (either extrapolated from historical data or estimated from scratch if the network is new), and dividing it up into one-hour slices (9:00–10:00, 10:00–11:00, etc.). The slice with the most erlangs of traffic is the peak busy hour. Bearer network dimensioning was, until the coming of optical fibre transmission systems, the dominant factor in network design, and is still interesting. The network designer may be able to choose between having a large number of small-capacity routes, or a small number of high-capacity routes. The choice is not straightforward. A large route can achieve higher average occupancy than several small ones. However, a large route may respond worse, in terms of blocking probability, to overload conditions. Now that transmission is so cheap, the cost of the call-processing systems is most important, and so call control dimensioning is paramount.

Flexibility more important than precision

The whole business of traffic forecasting and network sizing is getting more and more difficult, for the reasons noted above. The focus these days is on flexibility and speed of response, rather than trying to squeeze the maximum amount of traffic through what are usually pretty cheap assets. About the only part of the network where accurately sizing against customer demand is a key objective is the access network, where installation costs are high and service availability depends very visibly on network sizing.

Building the network

The network platform for a new service product is generally built speculatively, at least on a moderate scale. There is rarely time to secure a customer order before building the network, because speed of service activation is a key competitive advantage.

The network is built on the basis of work orders generated (increasingly electronically) from the designers' plans. However, when the network constructors (often contractors to the network operator) get out into the field, they commonly find that the situation on the ground does not allow the design to be executed exactly as intended. While major design changes are likely to be reviewed with the design team, the installation teams generally make small deviations from the plan without consultation.

This gives rise to one of the great problems of telecoms network management: when the installers depart from the approved design, they frequently do not update the plans to show what they have done. There immediately arises a difference in detail between the as-planned and as-built networks. My experience of working with many telcos is that most of them do not know, in fine detail, the configurations of their networks. This is a considerable handicap when it comes to diagnosing network problems.

Selling and provisioning

The selling of telecommunications services has much in common with other sales processes. The principal complicating factor is that it may be difficult for the salesforce to know in advance which services can be supplied to which customers. Service availability may depend, for example, on how close the network goes to the customer's premises, or on what equipment and network capacity is spare in that part of the network. Often a sale has to be referred to the planning department, which is costly and time consuming. Certainty of service availability is therefore an important, but difficult, goal of network platform design.

The selling of telecommunications services has much in common with other sales processes.

Provisioning means configuring an existing network so that it provides a particular service instance (i.e. a specific service for a specific customer at a specific network access point). Many service requests can be met by provisioning alone, without any need for additional network build. Indeed, a large proportion of service requests, in some operations as many as 90 per cent, are soft fulfillable. That is to say, the provisioning can be done electronically, without the need for manual intervention. Examples of soft-fulfillable service orders are:

- changing switch options, for example enabling access to premium rate services;
- changing tariff options, for example moving to a low-usage tariff;
- take-up of space on fractionally used media, as when the customer already has a fractionally provisioned E-1 circuit, and wishes to take up more of its capacity.[2]

Provisioning operations (often categorized as provide service, modify service, and cease service) can be a large part of a network operator's costs, as they are often the only part of the service lifecycle where human intervention is needed. Network operators are therefore strongly motivated to reduce the effort required. The first step in reducing the effort is one-touch or flow-through provisioning. This means that the service request is keyed in just once, by the network operator's staff, and the rest happens automatically. Still better is no-touch provisioning, where the customer interacts directly with the network operator's systems, via (for example) a DTMF phone, a WAP-enabled mobile or an HTML Web page.

Systems issues

For the software engineer, soft provisioning is challenging. What starts as a single simple user request has to be converted into often a large number of commands to different pieces of kit, distributed across a network, and presented in the various bizarre formats that the diverse network equipment expects. If something goes wrong with the provisioning

[2] Fractional E-1 is explained in Chapter 5.

of one network element, the operation has to be rolled back, putting all the equipment back into its original state. Since many provisioning operations are usually running concurrently, this can be excruciatingly difficult. Flow-through provisioning is therefore at present achieved only for a minority of network services.

Figure 10.3 presents an example of the sales and provisioning process for a leased line service.

The main steps in the process are as follows:

1. A business customer calls the network operator's customer care centre, and asks whether he can have a new E-1 leased line from his head office to his factory, how much it would cost, and when he could have it. The network operator has to respond quickly because its competitors will be after the business as well.
2. The customer care people send a service request, detailing what is wanted, to the network planning people. They look at their network inventory and topology systems, and probably also at some

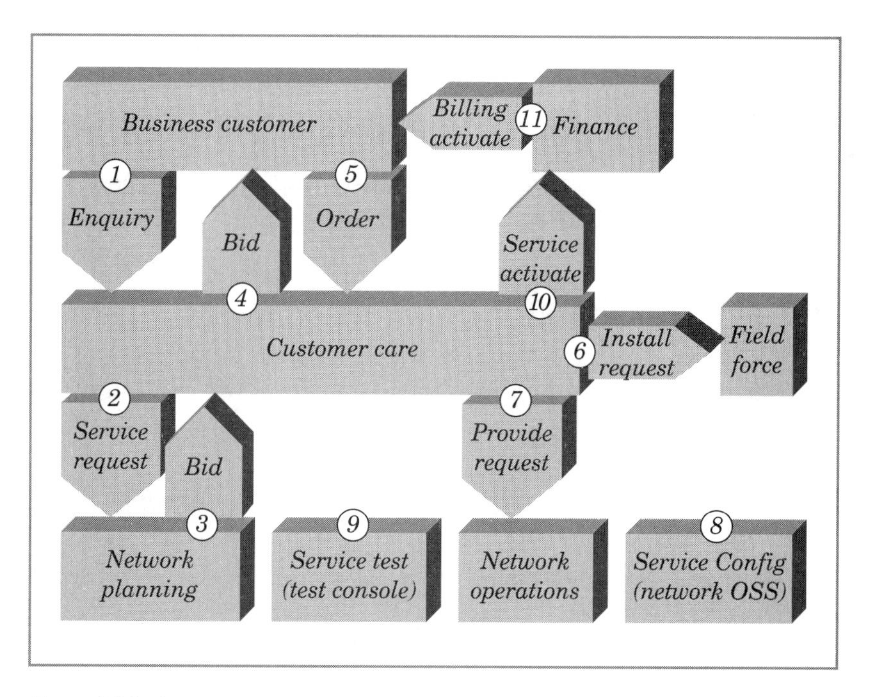

Figure 10.3 A provisioning process

geographical information systems (GIS) to establish what network capacity there is between the two locations.

3 They then cost up the necessary extension work, and pass back a bid to customer care.

4 Customer care presents a commercially wrapped version of the bid to the customer.

5 The customer is sufficiently impressed to place an order.

6 Customer care passes an installation request to the field force, who go to network nodes conveniently close to the required end points and (in this case) install LMDS WLL systems to reach the office and the factory.

7 A provisioning request is passed to the network operations team.

8 Network operations use their network *operations support systems* (OSS) to soft-provision the trunk network to complete the connection.

9 They then use the OSS to set up a loop-back test scenario, and use a test console to verify the operational quality of the circuit, and report back to customer care.

10 Customer care then asks finance to activate the service.

11 This means turning on the billing and advising the customer that the service is activated.

Operating the network

A network operations team typically looks after the following tasks:

- manually controlled soft provisioning (as we have just seen);
- preventive maintenance;
- surveillance, restoration and repair;
- tactical teletraffic management.

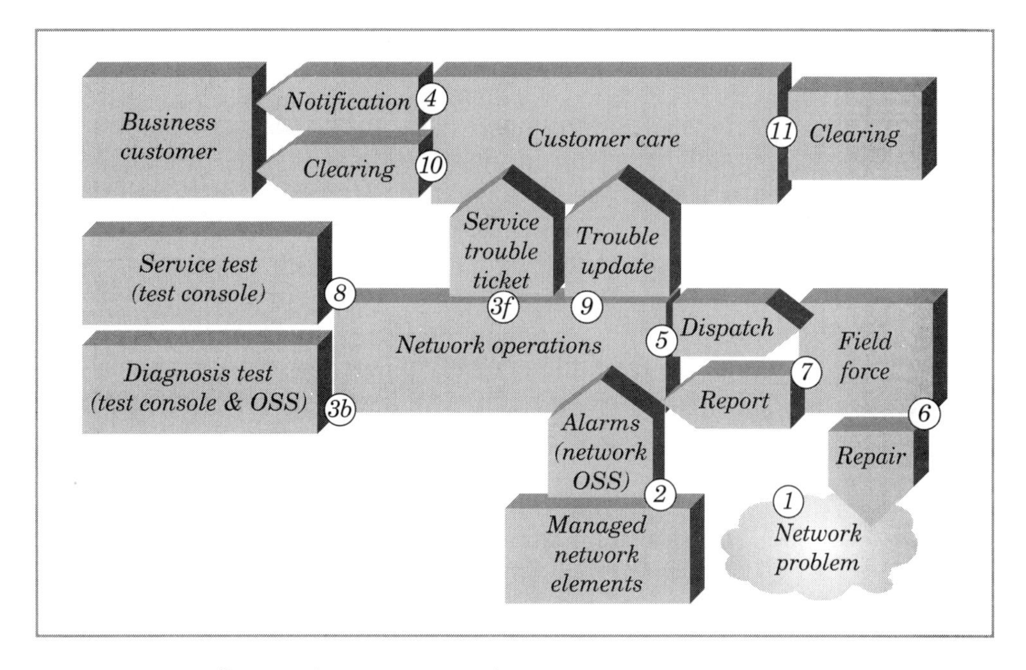

Figure 10.4 Surveillance, analysis, restoration and repair

Surveillance, restoration and repair

The processes of surveillance, restoration and repair are illustrated in Figure 10.4, which tells the story of what happens when there is a problem in the service that was provisioned in Figure 10.3.

1. Something goes wrong in the network.
2. The *network operations centre* (NOC) is monitoring the network by various means:
   - direct reporting through consoles connected to the major network elements;
   - indirect reporting through network management systems;
   - indirect reporting through relay contact management systems.
3. Because of this surveillance, the network fault leads to an alarm being raised in the NOC.

4 The NOC staff diagnose the problem. This is likely to involve a variety of processes and technologies:

- localizing the fault, which often will not be at the same location as the equipment that is complaining. For example, if the equipment at this customer's end of the LMDS link fails, the network end will report the failure;
- using diagnostic tools to investigate the problem;
- alarm flood filtering: often a single problem in the network will give rise to a host of alarms (sometimes even thousands of them). For example, if a high-rate SDH circuit is disrupted, all the SDH equipment through which all the SDH trails that the circuit carries will generate alarms, and so may the equipment where those trails terminate. The task of weeding through the masses of sympathetic alarms to find the one or two significant ones can be immense, and is a very desirable subject for automation;

- alarm correlation, which means looking at a number of alarms from different sources and spotting which ones are just manifestations of the same problem;
- root cause analysis (for example, in this case, looking at the pattern of alarms from the network end of the LMDS link, and determining that the root cause was a catastrophic failure of the user's end equipment, rather than, say, a gradual drift of antenna alignment);
- impact analysis, which means looking at the root cause and working out which network service instances and which customers have been affected, and looking at how the failure will affect their *service level agreement* (SLA). As a result of this analysis, one or more *service trouble tickets*[3] will be raised and passed to customer care;

–many of these tasks are difficult and require specialized knowledge of particular equipment. A common NOC

[3] Alias *fault dockets.*

staffing plan is to have a front office staffed with generalists, passing requests for detailed technical investigations to a back office staffed with specialists on the various types of equipment in the network. However, specialist staff are costly to train and retain, so increasingly network operators are seeking to hire the specialist skills from their equipment manufacturers as part of the equipment supply deal.

Specialist staff are costly to train and retain, so increasingly network operators are seeking to hire the specialist skills from their equipment manufacturers.

16 The customer care team notifies the customer (if they can contact him, that is) that the fault has been detected, and gives an estimated time to repair (which they may keep on updating, more or less credibly, until the fault is cleared).

17 Meanwhile, the NOC team has selected a restoration plan for the fault. Restoration means getting the service working again, which may or may not entail mending the broken equipment. For example, many networks have built-in redundancy in the form of restoration circuits, which can be patched in to restore service when a circuit fails. To reduce the demands on the expertise of the front-office staff, there is usually a list of common fault types, with prepared restoration plans for dealing with them. The reconfiguration of the network to achieve service restoration is yet another cause of the network planners' model of the network getting out of touch with its real state. In this example, the restoration will entail the actual repair of the equipment at the customer's site (the *customer premises equipment*). The NOC passes a work order to the *field force dispatcher*, who schedules a *field engineer* to attend to the problem. The field force scheduling problem is a computationally challenging one to automate; it needs to take account of the priorities of various requests, of staff work loads, of staff locations and travel times, of individuals' skills and equipment supplies. Often the field force will receive their instructions through mobile data terminals such as WAP phones or GSM-enabled laptops.

18 The field engineer somehow manages to repair the CPE.

19 He reports back to the dispatch centre, which passes the news back to the NOC.

20 The NOC staff test that the service really has been restored.

21 They update the service trouble ticket.

22 Customer care calls the customer to verify that the customer agrees.

23 Customer care clears the trouble ticket.

Tactical teletraffic management

Most large network operators have, in addition to their NOC(s), a tactical teletraffic management centre, which monitors and controls the network to ensure acceptable quality of service (particularly of blocking rates) in response to overload conditions (such as phone-ins) or network failures.

SPC circuit switches accumulate large volumes (often many kilobytes) of traffic statistics, which they send to their management centre every 5 or 15 minutes. Several specialized software tools address the storage, trend analysis and presentation of this data.

When an emerging traffic overload is detected, the operators may respond with a combination of expansive controls and protective controls. Expansive controls means adding more resources (for example, reserved trunks) to the network, to share the load. The most common protective control is call gapping. To appreciate how call gapping works, suppose there is an emergency of some kind, and a radio station gives out the relevant hospital's phone number. Hundreds of people start trying to call it. The hospital only has 10 lines, so most of the call attempts reach the exchange that the hospital is connected to, and then are blocked. That exchange becomes overloaded with trying to handle all the call attempts, and starts to fail to respond to requests from its neighbouring exchanges, or from its other local users. Call attempts between users entirely unconnected with the problem start to become blocked. The congestion can spread through the network, so that other exchanges also become overloaded.

Call gapping aims to prevent this sort of catastrophe. When the management centre spots the overload developing, it sends a command to all the ingress exchanges[4] in its network, telling them to put gapping onto the hospital's number. That means that each exchange will send only a maximum of x calls every minute to that number, and will block any others before they get into the network.

Meter and assure service

Quality of service measures

Quality of service is an important commercial issue for any telco in a mature market. Immature service markets tend to be oblivious of poor service quality, as witness the poor but accepted quality of early mobile networks, or of early VoIP. Once a market is mature enough for there to be serious competition, QoS often becomes as important as service pricing, in winning and retaining customers.

The ITU-T has developed a range of standard measures and objectives for quality of service. These relate mostly to POTS networks, and focus almost entirely on the QoS of the network. Network QoS elements include delay to dial tone, probability of blocking, or misrouting, or premature release, and call clarity. Network QoS however, is far from the whole story.

A wider view of QoS is provided by the content of the *service level agreements* or *service level guarantees* (SLGs) which network operators enter into with their larger customers, or which the regulators enforce on them. Among the non-network contributors to QoS are:

- the quality of the service management: is service provisioning, modification or cessation done correctly, quickly and at the agreed time?
- the quality of the billing: are the bills accurate and on time?

Measuring and assuring QoS can be very difficult for a network operator, because the network operator and the customer have quite different perspectives on the situation. The network

[4] That is, local exchanges plus any trunk exchanges that are connected to other networks.

operator has a network-centric view; he can find out a lot about the performance of all the network components, and even of his operational processes, but to synthesize all that information and relate it to a given customer is an enormous computational problem. The customer, on the other hand, is not troubled with all the internal details of the telco's operation, and can easily see and perhaps even measure the end-to-end service that he is receiving.

The network operator has a network-centric view.

ITU-T recommendations addressing quality of service

E.450	Facsimile quality of service on PSTN
E.800	Terms and definitions related to quality of service and network performance
E.845	Connection accessibility objective for the international telephone service
E.846	Accessibility for 64 kbps circuit switched international ISDN connection types
E.850	Connection retainability objective for the international telephone service
E.855	Connection integrity objective for the international telephone service
E.862	Dependability planning of telecommunication networks
E.880	Field data collection and evaluation on the performance of equipment, networks and services
X.140	General quality of service parameters for communication via public data networks

Service surveillance

Service level monitoring within a telco is a significant challenge. Broadly speaking there are three lines of approach, which are discussed in Chapter 11.

Bill and collect

Periodic pulse metering

Until recently the basis of billing information was *periodic pulse metering* (PPM). Each local exchange had a physical meter, very like a car odometer, for each customer line. The network generated high-frequency (50 kHz) meter pulses, at a slow rate for local calls (typically one every three minutes) and faster for long-distance calls. The meters counted the pulses. At the end of every billing period, a photograph would be taken of each exchange's rack of meters, then people with magnifying glasses would study the photographs and enter the numbers into the billing systems. Apart from being slow and costly to operate, the PPM system offered poor billing granularity (for example, a local call lasting one minute and one lasting two minutes would be indistinguishable). PPM, however, had the advantage that devices on the periphery of the network, for example PABXs and payphones, could process the meter pulses as their basis for user charging.

With the advent of all-digital networks, PPM has been very largely superseded by CDR-based billing.

CDR-based billing

Modern billing arrangements almost all depend on *call detail record* (CDR) processing, alias *automatic toll ticketing* (ATT)[5] or *automatic management accounting* (AMA). The idea is that every time a network node provides service, it generates a CDR, which identifies at least the end points of the call, its time of day and its duration (or data volume in packet mode networks). All the CDRs from all the nodes in a network are collated together electronically (a huge task) and then the aggregated CDRs are used to drive the billing process.

CDRs provide the extra information needed to produce itemized bills, and they enable fine-granularity, duration-based charging.

[5] A reminder of the days when call durations and distances were written down on cards by the manual exchange operators.

Cross-network charging

Where calls pass across more than one network, all the network operators need to be compensated. Where a network operator provides just a fixed circuit (leased line) to another, it is simply rented out. But where calls pass through a network's switches, usage-based compensation is required. A variety of arrangements for exchanging accounting information exist. The ITU has been particularly active in developing standards for cross-administration accounting across international boundaries (the ITU-T D-series recommendations). For mobile networks there are processes such as the GSM transferred account procedure (TAP), which works by passing files of CDRs from the roamed network to the caller's home network. Such inter-network exchanges can either be done by each network having a large number of transfer arrangements, one for each possible partner network, or through a single clearing house.

Generally, inter-operator accounting is subject to bulk discounting. This was a main driver for the removal of charging logic from the switches onto separate billing systems.

Tariff options

Network operators can charge for their services through combinations of:

- a charge for initial connection to the network and setting up the service;
- network access ('line') rental;
- usage charges.

Circuit switched local telephony may be free of usage charges, or charged per time. Circuit switched trunk telephony is charged per distance band, per time. Leased lines are charged per distance band per time (regardless of usage). Packet data services are either charged per time (regardless of distance), or according to data volume.

Fraud

Telecommunications fraud is in most cases a business process outside of the telco. However, fraud management is an important area of telecoms software engineering, so an outline of the fraud process will be useful.

Fraud is a major problem for many networks, accounting in some cases for losses of up to 20 per cent of a telco's proper revenues. Novelty is essential for effective fraud, so new forms of fraud are emerging all the time. The fraud models outlined below are therefore far from being an exhaustive list, but will still give a good idea of the range of scams possible.

Technology-independent fraud

Much telecoms fraud is not closely linked to the network technology, and relies simply on the fraudster being able to get credit from the telco. The idea of subscription fraud is to apply for connection to a network, perhaps presenting false identity information to do so, then resell the network service to third parties as busily as possible, until the telco ceases the fraudster's service because of non-payment. Where the network is a mobile one, it helps the fraudster to take the mobile phone abroad because the delay in charging for calls made in roamed networks extends the window of fraudulent opportunity.

Much telecoms fraud is not closely linked to the network technology, and relies simply on the fraudster being able to get credit from the telco.

PABX fraud

PABXs offer some special fraud opportunities. A PABX user may, for example, offer his associates cheap international calls by asking them to call him on his PABX extension, then using the PABX to dial the international number and then using the PABX transfer or conference feature to link the calls. Some PABXs offer *direct inward system access* (DISA), where an authorized user can call into the PABX, give a PIN, and get a secondary dial tone, from which long-distance calls can be made (at a cost to the caller of just the local call to the PABX). Such a PIN can command a significant value in some quarters.

Mobile fraud

With first-generation mobile networks, phone cloning is easy. To produce a clone of a first-generation phone the fraudster needs just two numbers: the phone's *electronic serial number* (ESN) and its user's *mobile identification number* (MIN). The phone's network will hold a table of valid ESN/MIN pairs as the basis of its user authentication. A clone which has a valid ESN/MIN pair is indistinguishable to the network from the original, and calls made on the clone will be billed to the customer who owns the original phone, until somebody spots the fraud.

Getting an ESN/MIN pair is not too difficult. TACS, AMPS and NMT all transmit them unencrypted. It is possible (for $100 or less) to buy scanners which listen out for ESN/MIN pairs. The fraudster gets one of these devices, and either stands on a bridge overlooking a congested main road (where everybody is turning on their phones), or operates a phone repair shop and uses his device there. Alternatively, the fraudster may be able to get access to the telco's database of ESN/MIN pairs. It is then simply a matter of blowing a new EPROM, installing it into a stolen phone body, and the clone phone is ready to sell, on a 'no-recurring charges' basis.

Wireline fraud

Wireline networks are exposed to particularly simple forms of fraud, which are popular in developing countries. The general idea is of *clip-on service access*; that is, taking a pair of crocodile clips and attaching another phone to an existing local loop, at some discreet spot. A more sophisticated form of wireline fraud, *phone phreaking*, is described in Chapter 6.

Calling card fraud

Possession of the card number and PIN for a calling card enables calls to be made as if one were the legitimate cardholder. One way to get these numbers is by loitering around call boxes and snooping over users' shoulders; the bobbing and weaving motion of the fraudster gives this its name of *shoulder surfing*. Another way is to find someone who has recently left a company which gave him a company calling card, and who bears the company a grudge.

Premium-rate services

Premium-rate services are not usually fraudulent in themselves, but they can provide the platform for a particularly elegant class of scam. The fraudster approaches the telco, acting as if he is a legitimate business person wanting to set up a premium rate service. The way such services work is that the telco charges the users, and then pays a proportion of the charges to the service provider. The service is set up, and the fraudster then makes as many calls as possible to the service, using stolen phones or other fraudulent network access. The telco pays him his first installment of revenues, and then he (or she) disappears before the scam is detected.

Staff fraud

Staff fraud is where somebody inside the telco, with privileged access to management systems, sets up a fraudulent operation. One example is unauthorized service provisioning where, for example, a local loop line is enabled without the corresponding activation of billing systems.

Fraud detection and control

All network operators of any size have a fraud detection and control operation. Fraud detection centres on detecting unusual patterns of network usage. Some well-known kinds of fraud can be addressed by specific detection systems. For example, mobile phone cloning can be addressed by velocity analysis. A system reviews all the calls made for each MIN, and compares their locations and times. If two calls on the same MIN are made so close together in time, and so far apart in space, that the necessary velocity of travel would be unbelievably great, then one or the other of them was probably made on a cloned phone. Other fraud detection systems use neural network techniques to detect usage patterns that are out of the ordinary in more subtle ways.

Once a fraud has been uncovered, the network operators often work closely with the police to monitor the fraudulent network traffic so that the fraudster can be caught red-handed. In mobile networks, the caller-tracing facilities described in Chapter 4 can be used to lead the police to the culprit.

Further reading

For a complete and well-accepted overview of what goes on inside a telco, *see* TeleManagement Forum (1999)(a), Flood (1997) sheds some light on the planning process, as does Cole (1999). Issues of service management are covered usefully by Hallows (1995).

11 Operations and business support systems

Most of the computer systems that we have reviewed so far are systems that are an integral part of the networks they belong to. If they were removed, the networks would stop working. This chapter introduces the other kind of telecoms computer system, the kind which supports the network, but which is not an integral part of it. Such computer systems divide roughly into two families:

- the ones that are intimately associated with the network technology, and which work (typically in real time) on detailed knowledge about the state of the network are called *operations support systems* (OSS), or *network management systems* (NMS);
- the ones that are largely independent of the network technology, which work on knowledge about the business's customers, tariffs and services are called *business support systems* (BSS).

Typically, the OSS are used by the operators' network operations teams, and the BSS by their finance and customer management people. In recent years service offerings have become complex, and service quality has become important competitively. As a result, the OSS and BSS in a typical telco may contain as much software as the actual network, and be just as critical to the business's success.

In recent years service offerings have become complex, and service quality has become important competitively.

The TMN problem

In the 1970s, telcos began to realize that management systems were going to be important, and that the management of a telco network was a significant challenge. Figure 2.2 shows how the idea of a telecommunications management network (TMN) arose. By the early 1980s, many telcos were experiencing serious difficulties with their management systems, and a 'TMN problem' was widely recognized. The problem was in two parts:

1. The equipment manufacturers were using OSS as a lever to exploit the telcos. Each manufacturer's network equipment had a different proprietary management interface, which only that manufacturer's OSS could work with. Telcos had to choose between two horrible alternatives:
   - a heterogeneous (multi-supplier) network, where there would be as many kinds of OSS as of network equipment; staff would have to learn each one's different user interface, and there was no hope of managing the network as a concerted whole;
   - a single-supplier network, where whichever supplier supplied the OSS was guaranteed a gravy-train of network equipment sales.
2. Each telco's network management was done differently, and there was no possibility of automating management interfaces between networks. So, for example, if telco A wanted to set up a circuit which traversed telco B's network, it had to be negotiated by expensive human beings. As deregulation in the 1980s hugely increased the number of telcos, this became more and more important.

Through the 1980s and 1990s a number of standards bodies wrestled with the TMN problem. Their work produced two quite different sorts of output:

- firstly, a universally understood language for describing TMN problems and network management processes;

- secondly, a set of standard interfaces for management systems and network equipment, to enable interoperation between manufacturers, and between networks.

As we shall see, only one of these outputs was much of a success.

The thought-scaffolding of TMN

The descriptive framework of the TMN standards (which originated from the ITU-T and the TMF) builds on the ISO OSI management model (ITU-T recommendation X.700). The framework is most useful as a fairly coarse descriptive tool, and much of its detail can be neglected for many purposes. The rudiments of the framework amount to three perspectives on the TMN: a *layer model*, a *functional area model*, and an *architecture*.

The TMN layer model

The layer model (Figure 11.1) views the TMN problem hierarchically, as five layers.[1] Each of the upper layers depends on the layer below, and the lower layers exist only to serve the layers above. The layer model is often presented as a five-layer pyramid, which has become a symbol of TMN thinking.

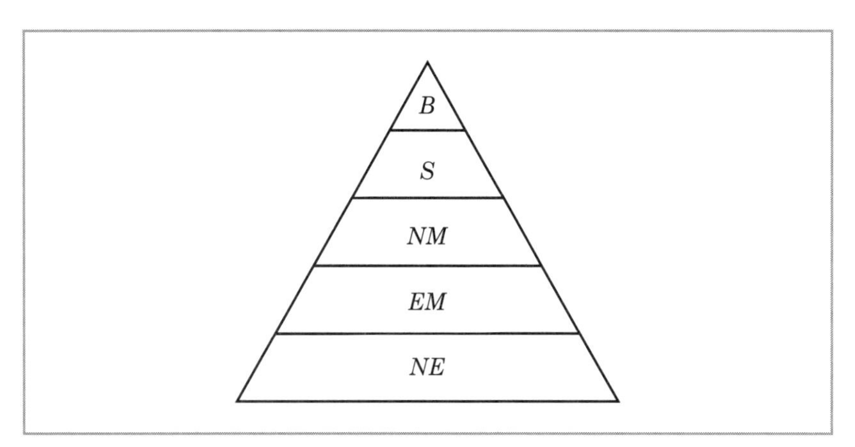

Figure 11.1 TMN layer model

[1] Some TMN writers have introduced a sixth layer, the *customer* layer, between the service and business layers. While this makes good sense, it has not been widely adopted.

The *network element* (NE) layer represents the network equipment, insofar as it is manageable. The *element management* (EM) layer is concerned with management work that is specific to particular kinds of NEs. *Element management systems* (EMS) therefore have a limited view of the network, but may have specialized knowledge about their NEs which does not need to be exposed at the *network management* (NM) layer. EMS are concerned with the maintenance and configuration of NEs, and with gathering operational statistics from them.

The network management layer is concerned with the network *as a whole*. By concerting the work of the EMS, an NMS can (the TMNthusiasts hope) effectively manage the end-to-end QoS, performance and connectivity of the network, without detailed knowledge of the NE technologies.

The *services* (S) layer holds the processes and systems that deliver services to customers and support the service life cycle. Typical service layer activities are service creation, order processing, provisioning, service assurance and billing.

The *business* (B) layer is concerned with how the offered services support the commercial purpose of the service provider.

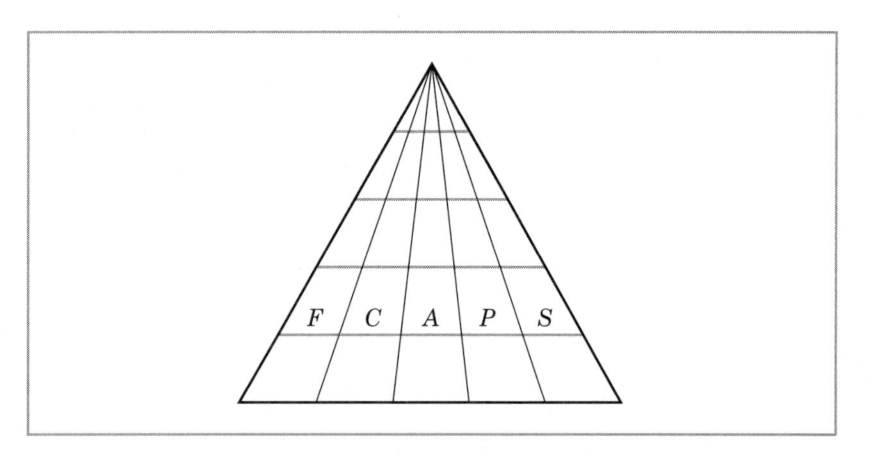

Figure 11.2 TMN functional areas

The TMN functional area model

The functional area model is another triangle, which can be overlaid onto the layer model, as shown in Figure 11.2. The vertical slices of this model are labelled FCAPS.

- F is for *fault management*: the detection, analysis and correction of abnormal operation;
- C is for *configuration management*: controlling and recording the state of configurable elements;
- A is for *accounting management*: enabling charges and costs to be determined and assigned;
- P is for *performance management*: evaluating the behaviour and effectiveness of the managed elements;
- S is for *security management*: generating, storing and distributing access control information.

The layer model and the functional area model can be used as a crude but widely understood way of locating a TMN problem. For example, a system may address fault management at the network management layer, or accounting management at the services layer.

The TMN reference architecture

ITU-T recommendation M.3010 defines a TMN systems architecture (Figure 11.3), which includes both functional blocks and reference points at their interfaces. The functional blocks are subdivided further in the ITU-T M.3000 series recommendations. In outline, the *operations system function* (OSF) corresponds to an OSS 'application' which monitors, co-ordinates or controls, for example, fault correlation or service order processing. The *workstation function* (WSF) represents the user interface of the OSS. The *mediation function* (MF) is concerned with conversion between data models at different levels of abstraction. An MF stores, adapts, filters, thresholds or condenses the data that pass through it. For example, an NM layer MF may receive information from an NE layer system and abstract it into terms of NM layer entities. The *Q-adapter function* (QAF) is a gateway, which converts between non-TMN-standard management protocols, and TMN-standard ones. The *network element function* (NEF) represents the TMN access point of the NE. The architecture assumes distribution of the function blocks across a data network, the *data communications function* (DCF).

Of the reference interfaces, several can be readily set aside.

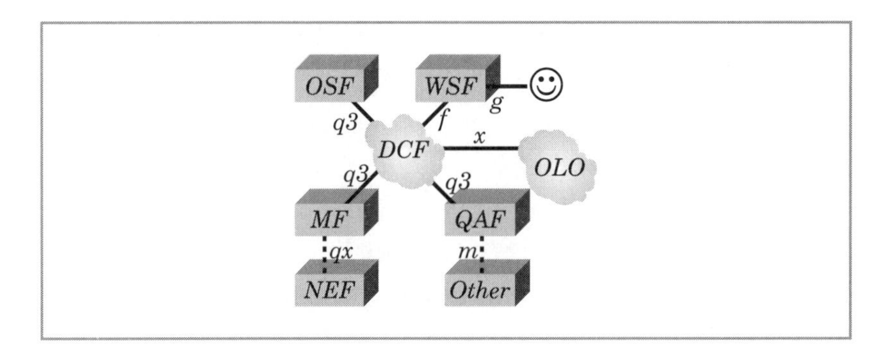

Figure 11.3 TMN architecture (after ITU-T M.3010)

The *g* reference point is defined as outside the scope of TMN standardization. The *m* reference point represents the management interface of a non-standard entity, and so is outside the scope of the standardization work. The *f* reference point has not been defined in any detail, and the *x* reference point is little further advanced. The two reference points that are of more substance are *qx* and *q3*.

The *q3* reference point is defined as a point where the management entities communicate using both standard protocols and standard information models; *qx* is defined as a point where the protocols are standard, but the information model may be nonstandard.

The possibility of systems supporting a *q3* interface has been enormously attractive to the telco TMN community. For the equipment manufacturers, who not only would have to implement it at some cost, but who also would have their arm-lock on the telcos broken by it, it has been less attractive.

Technologies and standards addressing the TMN problem

The TMN problem has been attacked simultaneously by a number of technology initiatives, of which the ITU-led TMN standards work is just one.

OMNM technology

Most of the technologies are implementations of *object modelled network management* (OMNM). The idea of OMNM is to represent each managed resource as a *management object*

(MO), which groups together all the relevant data and behaviour to model the resource for the purposes of management communications. Many aspects of how the resource works will be irrelevant to management communications, and so outside the scope of the MO.

A managed entity may present itself to its managing system as a collection of MOs, and the managing system may store a representation of the managed resources in terms of their MOs. Such a collection of MOs is called a *management information base* (MIB).

A particular MO is an *instance* of an object class. Its class definition (sometimes called *shared management knowledge*) may include:

- its attributes, or externally accessible data items;
- the management operations that it supports;
- its expected behaviour in response to management actions;
- the notifications (or events) that it may spontaneously send to its managing system.

MOs support the concept of *inheritance*, where a base class may be used to generate a number of derived classes, each of which includes all the properties of the base class. ISO has defined an MO class called *root*, from which all MO class definitions are meant to be ultimately derived. An MO class may be an abstract class (one which exists only to support the definition of another class by inheritance), or it may have one or more instances. An MO instance may *contain* MO instances (of the same or other classes). ISO has defined a container instance called *top*, which is meant to ultimately contain all MO instances. An instance may be identified by its full containment tree starting from *top*, that is, its full X.500 *distinguished name* (DN). Alternatively, it may be referenced from some other object that contains it, with a *relative distinguished name* (RDN).

MOs support the concept of inheritance, where a base class may be used to generate a number of derived classes.

Rationale

The rationale of OMNM is as follows:

- presenting the managed devices through object models conceals the detail of their workings from the managing system, which simplifies the managing system's job;
- OMNM enables standard MO classes to be defined, which will enable several implementations of the same generic device to use the same object model and be managed by the same system;
- where an equipment vendor wishes to add non-standard management features to a device, this can be done by deriving a specialized class from the standard one; thus the ordinary features of the device can be managed by any management system which supports the standard MO class.

Shortfalls

The OMNM approach has fallen short of these hopes. While standard MO classes were successfully defined for packet mode data communications equipment (because historically that kind of equipment was much simpler to model from a management point of view), the MO standardization efforts for the complex circuit mode devices in a telecommunications network were not very successful. The devices required large complex models, which were difficult to agree on, and there was a large legacy of pre-existing and complex management interfaces. While a number of standard MO definitions were produced for circuit-mode telecommunications equipment, they have not been implemented widely enough to make the management of multiple implementations from one management system a practical reality. Also, the management of specialized MO classes by systems which know only the standard MO proved to be impractical because the network operators typically wanted to be able to manage all the features of the device, not just the generic ones.

Standard methods

All the OMNM-based management technologies define standard 'methods' or operations, which their MOs are required to support. The standard methods are much the same in all of the technologies. They are:

- read attribute(s) from MO instance(s);
- traverse the MO containment hierarchy and discover what MO instances there are;
- write attribute(s) to MO instance(s);
- perform an action on MO instance(s);
- create MO instance(s) within a containment hierarchy;
- delete MO instance(s);
- notify the management system of a change in status (notifications are also called events, traps and alarms).

Managers and agents

Most of the technologies distinguish between the roles of a management *agent* (the management entity in the thing that is being managed) and a *manager* (the management entity in the thing that is doing the managing). An agent is generally passive, receiving requests from the manager and issuing responses to them, while the manager is generally active, issuing requests and receiving responses. The exception is the *notify* operation; here, the agent issues the notification off its own bat, and the manager responds to it.

'TMN' technology

We have seen that 'TMN' is firstly a framework for describing management problems. Secondly, 'TMN' is a collection of technologies developed and collated by the telco community in an attempt to address TMN problems. TMN technology standardization has been led by the ITU-T and the TeleManagement Forum (TMF, until 1999 the Network Management Forum, NMF).

The chief focus of these bodies' efforts has been to provide standard interfaces for the q3 reference point. The ITU-T's efforts are documented in the M.3000 series of recommendations; the TMF's in document sets called OmniPoints.

TMN interface protocols

To start with, the OSI *common management information protocol* (CMIP) was selected as the interface protocol, with FTAM (the OSI file transfer protocol) also permitted for some bulk data-transfer requirements.

CMIP is an application layer protocol defined in ITU-T recommendation X.711 (ISO/IEC 9596-1). It provides a set of application services, which are documented separately as the *common management information service element* (CMISE: X.710 or ISO/IEC 9595). The operations supported include all those listed above, and mostly their implementation is done very thoroughly. For example, a single CMIP M-SET operation using the *scope* and *filter* features could adjust an attribute of every MO instance in the network that meets some selected criteria. CMIP allows for transfers of large and complex blocks of data in single operations, and also enables a management agent to send a whole string of linked replies to a single request. The technical weaknesses of CMIP are that its discovery capability is limited, that it is very, very, complex, and that it has so many optional features that two implementations of the protocol are unlikely to interwork satisfactorily.

CMIP uses two other OSI application layer elements. The *association control service element* (ACSE, ITU-T recommendation X.217) provides an application context analogous to a TCP/IP pipe. CMIP and ACSE both use the *remote operations service element* (ROSE, ITU-T recommendation X.219), which provides a service analogous to remote procedure call (RPC).

At the presentation layer, the CMIP stack uses ASN.1 and the associated *basic encoding rules* (described in Chapter 12). At the session layer and below, it was originally intended to use the connection-oriented OSI stack. In real life, many CMIP implementations use TCP/IP (RFC 1189) instead. *CMIP over TCP/IP* is abbreviated to CMOT.

Above the protocols

Above CMIS, the ITU-T defined a whole raft of *systems management functions*, which were intended to be universal functions for incorporation in many or all MO classes. For example, the *state management function* says that classes should include an *operational state* attribute, which says whether the object is disabled or enabled, and an *administrative state* attribute, which says whether it is in or out of service.

Some ITU-T systems management functions

X.730 Object management function

X.731 State management function

X.732 Attributes for representing relationships

X.733 Alarm reporting function

X.734 Event management function

X.735 Log control function

X.736 Security alarm reporting function

X.740 Security audit trail function

GDMO

The ITU-T standard for defining an MO class is X.722 (ISO/IEC 10165) *guidelines for the definition of managed objects* (GDMO). GDMO defines how one should specify a class's:

- name bindings (that is, its containment);
- attributes and their types;
- notifications;
- inheritance;
- behaviour.

Most of these are defined using ASN.1. However, the behaviour is defined in unstructured English. GDMO is horribly difficult to work with, and has suffered from being a telecoms-specific object-definition technology at a time when object modelling has found widespread use in mainstream IT. GDMO is therefore being superseded, even in the TMF, with UML and XML.

The twilight of prescriptive TMN

All these TMN technology standards have in general been poorly taken up by the equipment manufacturers. While the manufacturers have been keen to oblige the telcos, they have equally been aware of the cost of implementing the complex TMN standards, of the poor fit of some of the standards to real-life problems, and of the weakening of their commercial position which a widely implemented uniform management interface could bring. Consequently, management based on the CMIP protocol and standard class definitions has never reached critical mass. So, while the descriptive model of TMN remains valuable, the prescriptive standards that accompanied it have been less of a success.

Another angle on the problem of prescriptive TMN is presented in Figure 11.4. The TMN standards really addressed the *assure*, *operate*, *provision* and *build* business processes, all of which are cash drains. TMN technology had little to offer the *sell*, *bill* and *collect* processes, which are the ones that bring in a telco's income. Therefore the funding for TMN systems within the telcos has always been limited, and so the return on investment for an equipment manufacturer developing TMN standard management interfaces has been poor.

The TMF has moved on from its former prescriptive role to that of a facilitator. Its Smart TMN programme has brought together several consortia of telcos and equipment vendors to build advanced interoperable management demonstrators, with a view to developing standards after the implementations have been proved. TMF has also moved rapidly to accommodate grass roots trends, notably the acceptance of SNMP and, more recently, CORBA, as more likely alternatives to CMIP.

> Recognizing that the TMN technology would be a long while coming, the Internet community went ahead and developed its own network management framework.

Internet management technology

While the TMN community was thrashing away trying to address the TMN problem very thoroughly, the Internet community took a more nimble approach. Recognizing in the mid-1980s that the TMN technology would be a long while coming, the Internet community went ahead and developed its own network management framework. Initially this addressed only fairly simple IP devices, in smallish networks that had no very onerous availability requirements. Because the internet management framework was simple and easy to implement, it gained widespread acceptance. Now that internet technologies have come to dominate the future of telecommunications, the internet management framework provides by far the most widely implemented (at least in terms of numbers of devices which support it) system of management standards.

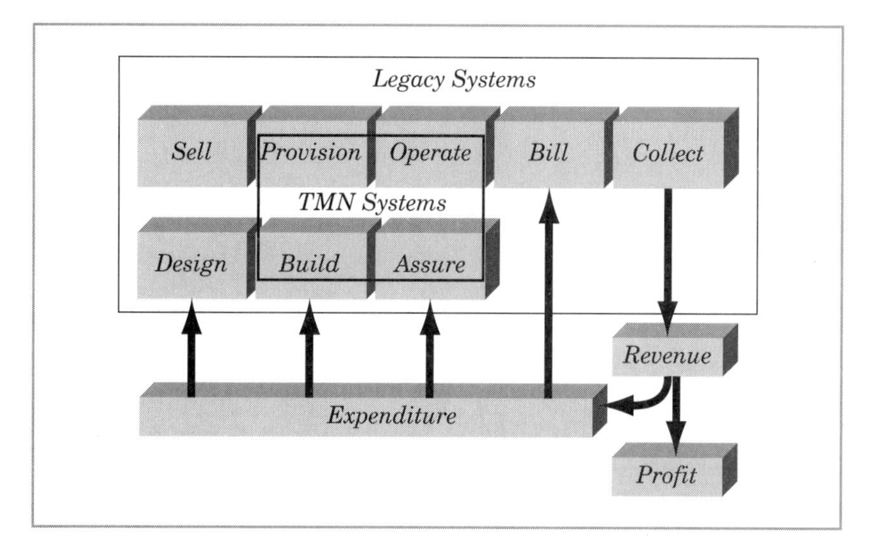

Figure 11.4 TMN technology and cash flows

SNMP

The basic protocol for internet management is called 'a *simple network management protocol*' (SNMP, defined in RFC 1157). SNMP was conceived as a simplified version of CMIP, and so its message types are closely aligned:

- Get-Request;
- Get-Next-Request;
- Get-Response;

- Set-Request;
- Trap.

SNMP runs over UDP/IP.

SNMP MIB definition

The internet rules for defining objects are documented in RFC 1155 '*structure of management information*' (SMI) and RFC 1212 '*concise MIB definition*'.

The SMI, like GDMO, is based on ASN.1, but with a very restricted subset of the data types that ASN.1 offers:

- integer;
- octet string;
- object identifier;
- null;
- sequence;
- sequence of.

RFC 1213 defines 'MIB I', a collection of object definitions which were sufficiently simple and universal to find very widespread implementation. RFC 1215 defines 'a convention for defining traps for use with SNMP', and also a set of seven generic traps: coldStart, warmStart, linkDown, linkUp, authenticationFailure, egpNeighbourLoss and enterpriseSpecific.

Real-life SNMP implementations usually involve the basic standard MIB and traps, plus other standard or enterprise-specific MIB extensions and traps. Thousands of enterprise-specific MIBs have been developed by equipment manufacturers. Therefore a valuable feature of a network management system is a 'MIB loader', which enables the system to consume an SMI file of an enterprise MIB, and automatically adapt the management applications to it.

Also, a number of standard specialized MIBs have been agreed. For example, RMON (RFC 1757 '*remote network monitoring MIB*', and RFC 2021 '*remote network monitoring MIB* version 2 using SMIv2') supports the monitoring of packet throughput, per port and per address pair, and provides traps which an RMON agent can use to warn of traffic overloads. SMON takes the idea further and monitors the performance of the whole switch, rather than single ports, and supports the analysis of performance against IEEE 802.1D service priority tags.

Limitations of SNMP 1

The first version of SNMP (now called SNMP 1) had several limitations. It was made simple to implement in the network devices, at the cost of demanding a lot of network traffic and a lot of work from the management system. Because UDP/IP is a connectionless protocol without confirmation of delivery, traps can get lost in the network. Traps are elicited from the devices by polling every device every 30 seconds or so. The network load created by the polling grows with the size of network, and limits the size of network that a single SNMP manager can practically manage.

SNMP uses the bearer network (IP) for management communications. While this saves the expense of constructing a separate management network, it means that if the bearer network fails (a time when network management becomes very important), the management systems can no longer manage the devices.

SNMP's authentication arrangements are very limited, so it is difficult to keep control over which management system is fiddling about with which devices. Lastly, SNMP's management model is strongly agent-manager, and peer-to-peer communications between managers is not addressed.

SNMP 1 MIB II

In the late 1980s and early 1990s, there were attempts to define an SNMP 2 addressing some of these limitations. However, different groups drafted a number of different SNMP 2 RFCs. This diversity, plus a continuing weakness in SNMP security features, meant that SNMP 2 gained little ground. At about the same time, a more elaborate MIB, MIB II, was developed, and was quite widely implemented, using the SNMP (1) protocol. This combination of SNMP 1 and MIB II is sometimes called SNMP 1.5.

SNMP 3

RFC 1157 (mentioned above) introduces Version 3 of the Internet management framework, known as SNMP 3. SNMP 3 uses Version 2 of the simple network management protocol (SNMPv2, defined in RFC 1905), and also SMIv2.

SNMPv2 has more effective security and access control mechanisms. SMIv2 introduces a more rigorously object-modelled MIB structure, and a number of additional data types:

- Integer32;
- enumerated integers;
- Gauge32,;
- Counter32;
- Counter64;
- TimeTicks;
- IpAddress;
- Opaque;
- BITS.

Policy-based networks

A quite different approach to network management has been developed under this name. Or to be more exact, at least two approaches.

The more developed one is policy-based capacity management. The idea is that IP network devices need to be able to allocate appropriate capacity to each connection, according to its business value or other status. In policy-based capacity management, a network service operator has a policy decision point (PDP), which can identify, for each connection endpoint (IP address) and service, the appropriate quality of service required. When a user requests a service (possibly over a quite different part of the Internet), the device which sets up the connection (a *policy enforcement point* or PEP) queries the appropriate PDP using the *common open policy service* (COPS). COPS is an emerging technology, and has yet to be fully standardized or widely implemented.

The other approach that calls itself 'policy based' is an attack on the problems of fault management. The idea is that a policy is a rule which resides in a network management system, which says 'if the following set of things goes wrong, then do such and such to restore the network'. In other words, it is about automating restoration plans. While this is a challenging problem, it is essentially internal to the management system, and so does not require the definition of inter-system standards.

Web-based management

Yet another approach to network management interfaces is to exploit the universal remote access and user interface capability of the World Wide Web. HTML is easy to implement, well known, and user-readable, so a single HTML interface could support both remote management systems and human operators. However, HTML is designed to support presentation for human understanding, not for machine interpretation. Also, http is a pull-based protocol, and therefore does not readily support traps.

The Web-based enterprise management forum (WBEM) is promoting standards for Web-based management, including:

- *hypermedia management schema* (HMMS), a data description schema for management information;
- *hypermedia object manager* (HMOM), a common class model;
- *hypermedia management protocol* (HMMP), which carries HMMS over http.

TL 1

Yet another approach to the standardization of management interfaces was taken, very successfully, by Telecordia (then called Bellcore). When the Bell operating companies were procuring their Sonet networks, they insisted, with their immense commercial weight, that all the equipment should be compatible with their management system. The standard interface was *transaction language 1* (TL 1), defined in Telecordia TA-396. TL 1 is a pragmatic subset of the ITU's MML standards (from the ITU-T Z.300 series). It is ASCII-based, and

Another approach to network management interfaces is to exploit the universal remote access and user interface capability of the World Wide Web.

suitable for both management systems and craft terminal access.

The guaranteed initial market for TL 1 implementations, together with its telco orientation, made it a very popular standard, and it is still one of the most widely implemented (if unglamorous) management interfaces.

Relay contacts

Alongside all these computer-oriented approaches to management interfaces, a far older and simpler technology has persisted: relay contacts. Many network devices provide pairs of contacts, which change state when something goes wrong. By wiring all the contact pairs back to a control centre, a crude picture of the status of the equipment can be developed.

While the amount of information that relay contacts provide is small, their extreme simplicity and widespread availability mean that they have continued to be implemented into the new millennium.

TINA

In a parallel endeavour to the TMN standards work, the Telecommunications Information Network Architecture Consortium (TINA-C) developed still another approach to management systems standardization. The TINA approach is based on CORBA as its distribution mechanism, and is outlined in 'Overall concepts and principles of TINA', V 1.0 1995, TINA-C.

Practical telco network management architectures

Figure 11.5 illustrates a typical OSS architecture for a national telco in 2000. The network elements are managed by a collection of technology-specific EMS, mostly via ad hoc proprietary protocols, but also some via TL 1 and SNMP. A few of the network technologies also have their own network-level management systems, but these address only the particular technology concerned.

There is an NMS, which does fault management integration, some performance management integration, and possibly some electronic network provisioning. This system interfaces to the EMS via predominantly SNMP or proprietary ASCII interfaces, but also some CORBA and CMIP.

Integration of the NMS with other business systems is patchy. Usually it is integrated with a trouble management (workflow) system. Integration with order handling, workforce management and planning systems is much less common. Notably, billing information still usually passes direct from the switches to the billing systems, or via a billing mediation system. It is as if billing information is just too valuable to entrust to those network management people!

Network management systems

EMS and NMS are usually built with similar software technology, and offer similar user features, but at different levels of abstraction. The following sections outline typical EMS/NMS features, under the FCAPS headings.

Fault management

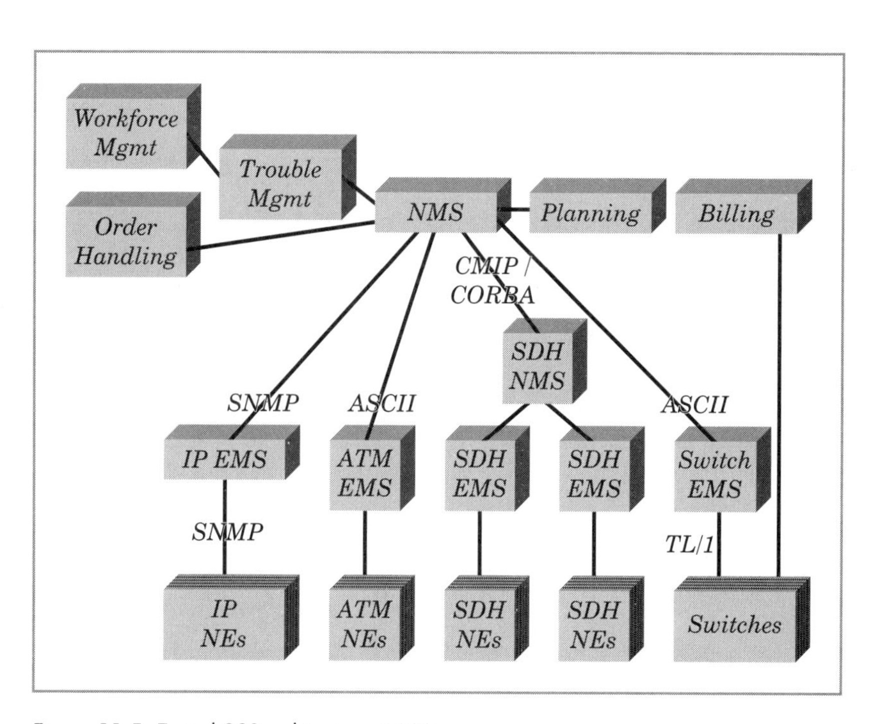

Figure 11.5 Typical OSS architecture, 2000

Schematics and geoschematics

Usually, when an item has an alarm associated with it, it changes colour to alert the operator.

Schematic displays show the managed network logically, without attempting to present its physical layout. *Geoschematic* displays present the managed network in a style that is halfway between a schematic diagram and a geographical map. While not geographically exact, they give a good idea of roughly where each item is.

Schematics and geoschematics are usually layered, so that a top level display may show a single icon to represent a large switching site, a lower level display will show the detail of the site, and a lower level still will show the configuration of the cards in the switches. Usually,

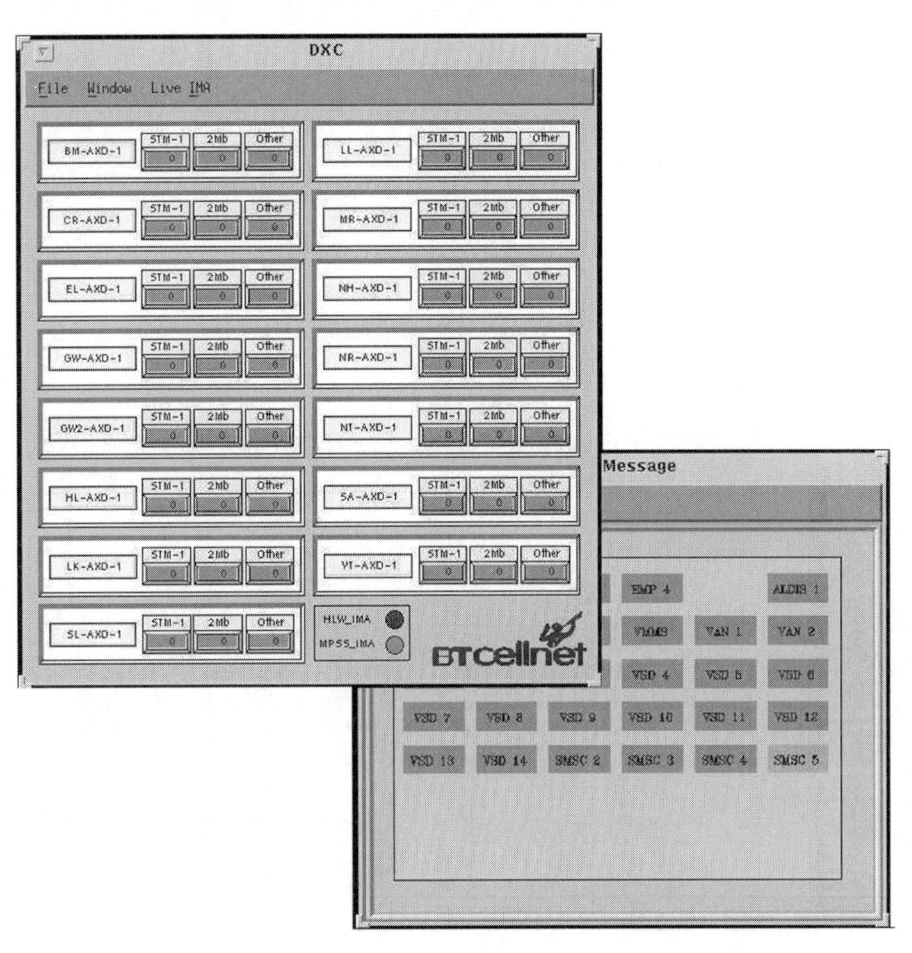

Figure 11.6 Schematic display

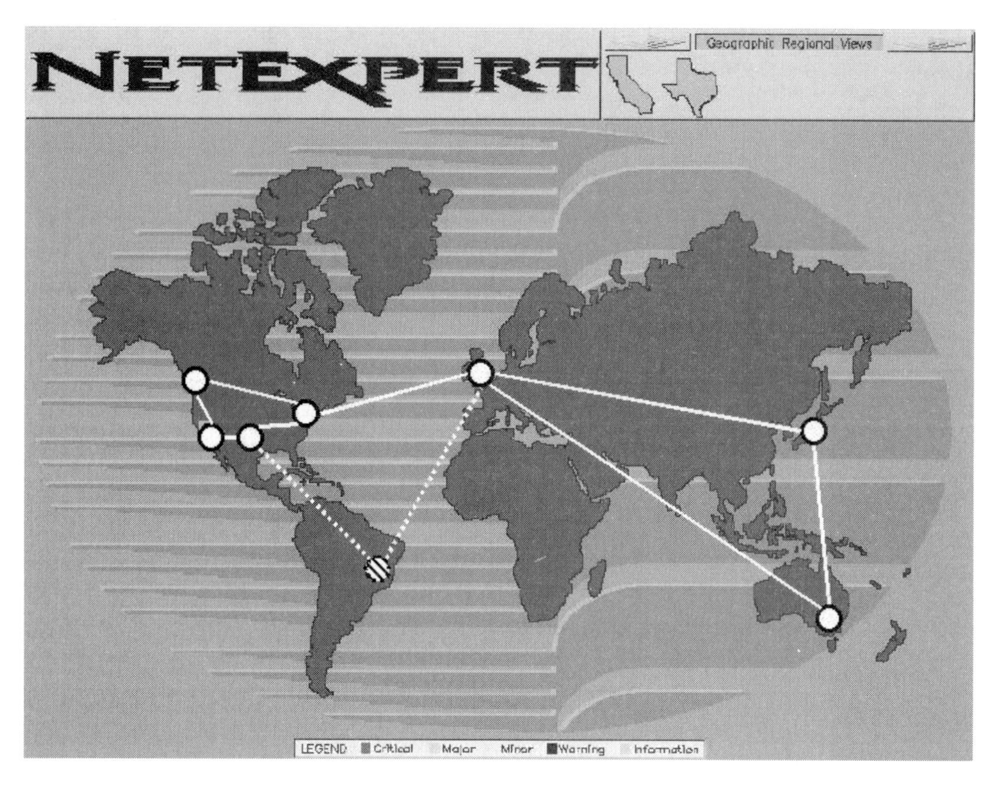

Figure 11.7 Geoschematic display

when an item has an alarm associated with it, it changes colour to alert the operator. 'Bubble-up' means that if an item on a lower-level view changes colour to warn of an alarm, the corresponding item on the next level up changes colour as well, and so on up to the top level. If an operator wants to move down through the display hierarchy, this can be done by 'drilling down' (typically clicking on the icons).

Figure 11.6 is an example of a schematic display from a network management integration system that BT Cellnet uses to manage its mobile network. The lower panel presents the overall status of several devices at a site; the upper panel shows the status of one device, in greater detail.

Figure 11.7 shows an example of a geoschematic display, presenting the status of a number of sites across a territory. When using a display that presents individual network devices, it is often possible to 'cut through' to a terminal emulation (Telnet or VT220) session connected to the device selected. While this requires the system operators to be familiar with the devices' craft terminal interfaces, it can save immense expenditure on management interface software.

Alarm windows

The other most common display feature is an alarm window. Alarm windows present a list of alarms, each alarm appearing as a horizontal row of information, as illustrated in Figure 11.8. Usually the operator will have some control over which alarms are let through to appear on the display (filtering), and what order they appear in (time of arrival, time of generation, priority, etc.). Each alarm row will include at least the identity of the device that raised the alarm, the time of generation and the alarm type. Usually a lot of detailed information is also presented, and the row is colour-coded to indicate its priority.

Many alarm windows support multi-user operation by allowing an operator to acknowledge an alarm (to show he or she is working on it), and then marking it so that all the other operators can see that it has been acknowledged. Another common feature of the more up-market alarm windows is the facility for an operator to automatically have a trouble ticket generated (in a linked workflow system), and have the alarm and the trouble ticket tracking each other's status.

Alert Display Window : 1

Display Alert Management Trouble Ticket Device Mgmt Quit

AMO	Description	Count	Create Time
TB24_0528E(Ashton-under-lyne)	TGP59 DROP 143 001 000 DURATI	1	09:24:22
TB24_0528T(Ashton-under-lyne)	CELL 028 000 000 INS OO	1	09:23:00
TB07_0211T(Milnthorpe)	CELL 005 000 000 INS OO	1	09:19:29
TB01_1239T(Thurso)	3MEG 045 001 000 DURATION ALA	1	09:13:52
TB12_0167U(Leamington_spa)	CHGR 047 001 000 INS IN	1	09:06:07
TB26_0182U(Buckingham)	3MEG 006 001 000 DURATION ALA	1	08:47:16
TB19_1231T(Corwen)	CHGR 035 002 000 INS IN	1	08:44:18
TB19_1300U(Madley)	CELL 026 000 000 INS OO	1	08:10:53
TB29_0079U(Membury)	CHGR 007 004 000 INS IN	1	08:04:26
TB02_0031U(Bloxwich)	CHGR 002 001 000 INS IN	1	06:13:09
TB02_3107U(Big_peg)	CHGR 086 003 000 INS IN	1	06:12:52
TB16_0690U(Ashford)	TGP25 LINK 448 000 000 DURATI	1	05:41:44
TB25_0315T(Liverpool_city)	P/A 048 002 001 INS-ACT IN	1	04:27:28
TB26_0367E(Streatley)	TGP32 DROP 131 001 000 DURATI	1	04:24:02
TB07_0118U(Sunderland)	CELL 020 000 000 INS OO	1	04:01:42
TB12_0288T(Gainsborough)	CHGR 017 002 000 INS IN	1	03:31:04
TB15_0838U(Alsager)	SCC 066 002 000 INS-ACT OO	1	03:00:31
TB26_0767T(Banbury)	MSCP 036 001 000 INS-ACT OO	1	02:19:25
TB16_0733T(Eastbourne)	TGP26 LINK 231 000 000 DURATI	1	02:17:27

Figure 11.8 Alarm window[2]

[2] For security reasons, the screen shots in Figures 11.6, 11.7 and 11.8 are based on simulation data, not live network traffic.

Routining

Routing is a very specific fault-management function which is deployed in most wireline local loops. A wireline local loop has an impedance which is a function of its length. For any particular local loop, it is possible to determine upper and lower bounds for its normal impedance. Routining is a process where a local exchange switch goes round all of the local loop connections in turn, applying a test voltage and measuring the loop impedance. If the impedance is out of range, an alarm can be raised and the network operator's engineers can investigate and remedy the problem, often before the customer realizes that there has been a fault. As switch connections migrate from direct local loop connection to TDM connection, the routining function is taken over by specialized loop test equipment outside the switch. Wireline customers may notice that sometimes on a quiet evening their telephone makes a little click or ping as the routining voltage is applied.

If the impedance is out of range, an alarm can be raised and the network operator's engineers can investigate

Configuration management

A management system's configuration management (and to a lesser extent fault management) capability depends critically upon how much information it stores about its managed network. Some systems store a model of not only what devices are in the network, but what each device's last known status and configuration was. Some systems remember what devices are there, but nothing about their state. Other systems (called MIB-less) do not even bother to store a list of the devices in the network.

Where a full network model is stored, the operators can edit changes to the network configuration image in the NMS, then download them into the managed devices. Where this is not available, the most common alternative is configuration management by cut-though. While this looks crude, it can often be justified; the software cost of complex configuration management facilities is high, and often a network device's configuration can be left unchanged for most of its life.

A persistent problem in telcos is that the management systems lose track of what the actual network configuration is. Therefore many management systems offer autodiscovery features, which poll every known device in the network, asking it about its neighbours, and so construct an up-to-date network map. Autodiscovery is particularly well supported by SNMP.

Getting a management system to do provisioning is generally a major challenge. What starts as a simple provisioning order, expressed in terms of network facilities and telephone numbers, has to be converted into the internal data models and obscure interface languages of the network elements concerned, then distributed to them. Even working out which are the right elements is a challenge. There are also difficulties in synchronizing the changes, and in rolling back a complex series of changes if one of them fails.

Accounting management

Until the late 1990s, almost all telco networks were circuit switched. Accounting management was done through CDR processing, and the CDRs were not handled by the general network management systems at all. Now that many telco networks are IP based, there is a new opportunity for usage information to be integrated into the main stream of network management.

Performance management

Network performance management is often done in specialized performance management systems. This is because of the huge data volumes that have to be processed, and the requirement to process them in near-real time. A number of telco network performance management tools have been developed, to collect, collate, store and present the performance information. Regulators often require telcos to be able to prove things about their historical performance, and so the performance data has to be archived for several months.

Security management

Security management, like accounting management, has not historically been a major part of the management systems world. In circuit switched networks, where the management network is distinct from the bearer and signalling networks, control of access to the network devices' management features could be achieved simply by controlling access to the management systems' operator consoles. Now that customers are offered windows into the telcos' management systems (customer network management, or CNM) and IP networks make remote access to management features easier, security management is becoming more of a challenge, and is one of the attractions of SNMP 3.

Service management systems

Service management systems are a fairly new thing in telecommunications. Historically, the problems of network level management have been so absorbing that only recently have many telcos had the solid base of integrated network management on which to base service management.

Service surveillance

The area of service management that has yielded most success so far is *service level surveillance*. The surveillance data that network management systems offer is stated in terms of equipment failures, and possibly circuit failures. What telcos increasingly need is information stated in terms of services and customers, or of SLAs.

Without such information, it is difficult for the telco to prioritize restoration work. If two items of equipment fail, which one should be fixed first? To work out manually which customers are affected, and how, by each failure, is usually so difficult that nobody even tries. Restoration is often done on a first-come, first-served basis without regard for how this relates to the telco's service obligations. Without service status information, it is difficult for the telco to have a fair basis for service quality reviews with the customers. Often it is easy for the customers to see what the service status is because they see the service end to end, as a whole. For the telco, which sees a plethora of network devices and connections, the state of the customers' services is usually difficult to establish.

Three approaches to service level surveillance are emerging:

- perhaps the more thorough approach is to take existing information about the status of all the things that contribute to service performance and integrate the information. Most surveillance systems address only the network or equipment levels; to get from these to service level monitoring entails the integration of information from many sources, and then its abstraction to the service level, and perhaps from there on to SLA automation. This is technically difficult, and also relies on the network operator having reasonably good information about how all its network is configured, and how particular network elements relate to particular customers' services. However, it can be done, and when it can, it yields results which are demonstrably precise. While this approach has the advantage of offering precise results, on which SLAs can be readily based, it depends on the above mappings being known. In IP networks, in particular, the relationship between the network topology and the paths taken by traffic changes dynamically, and so this approach becomes unworkable;

Most surveillance systems address only the network or equipment levels.

- in such cases, the second approach is more successful. It is to use test systems to generate service-usage traffic, and measure samples of the end-to-end service performance of the network. From these samples, a statistical view of the status of other similar services can be derived. While this approach is robust against varying or incomplete network-service mappings, its results are only statistically valid (at best), and a large volume of simulated traffic may have to be used if statistically significant results are wanted. Also, even if the results of the simulation can be made statistically representative, they are difficult to make sufficiently watertight to base an SLA on. And this approach can address only the network contributors to QoS, whereas the approach above can integrate things such as help desk response time;

- lastly, service surveillance can be based on network usage records such as CDRs. While these are a ready-made source of accurate service information, they are limited in only describing the events where the customers received sufficiently good service for a CDR to be generated. So where, for example, a mobile user is unable to get an access connection into the network, they fail to represent the problem.

Billing systems

CDR generation

CDRs, AMA or ATT records[3] (introduced in Chapter 10) are typically generated from the circuit switches through which the billable calls pass. As each stored program control (SPC) circuit switch has its own well-developed idiosyncrasies, CDR formats tend to vary from one switch to another. However, they generally include at least the following fields:

- calling number;
- called number;
- number the call was delivered to;
- call start timestamp;
- call end timestamp;
- call termination reason.

As IP networks become widespread in telcos, the desire for usage-based billing of IP services is leading to the addition of CDR generation capability to packet mode devices. In these cases, the CDRs also include data throughput information (in packets or octets).

CDRs can also be generated by network surveillance systems. For example, there are commercial products which intercept the SS7 traffic around the network, develop a model of the service traffic, and generate CDRs accordingly. Such systems are particularly attractive where a network operator wishes to multisource its switching equipment, because they remove the need to work with multiple CDR formats.

CDR collation

In a network with, say, 10 million customers, there may be 100 million calls per day. Each call may generate more than one CDR (depending on how smart the switches are about co-ordinating their CDR generation), so each day there may be,

[3] 'CDR' is used here to cover all three terms (which are all functionally the same thing).

say, 300 million CDRs to process. The CDRs need to be sorted through so that all the information relating to each call can be brought together into a single 'collated' or 'aggregate' CDR. This is not at all a trivial problem.

Billing system elements

Figure 11.9 shows the billing system elements. CDRs are typically produced by the switches as data files, which are transferred to the billing systems by FTP or FTAM, at an interval which varies between a few minutes and 24 hours. Often the CDRs' first destination is a mediation system, which is responsible for transporting[4] all the network's CDRs back to the billing site, and for any format conversions necessary. Following CDR aggregation, the CDRs are rated. This means assessing the time of day, source and destination numbers to determine the charge band, and adding a provisional charge to the CDR, according to the current rating rules. Periodically the billing system generates bills, applying further business rules to give high-usage discounts.

Given the rapidity with which new services are introduced, and the convergence of telecoms with other business areas, flexibility and versatility are important in billing systems. One useful approach is event-based billing, where the billing system can be configured, without code changes, to bill for many kinds of chargeable events. Thus a single system may do the billing for voice calls, Internet access, pay-per-view and directory services, and can offer cross-product discounts.

Pre-paid services

In market areas where credit management is unattractive either for the telco or for the customers, pre-paid services have become popular. Each user buys an amount of credit, which the network decrements when the service is used (and often also periodically, as a form of connection rental). The users can add to their credit balance either by requesting a charge to their credit card, or by buying vouchers and using unique numbers on the vouchers to prove it to the network.

There are three main ways of implementing pre-paid ser-

[4] There are still parts of the world where this is done manually, with tapes and vehicles.

vices. One is to use an ordinary billing architecture as described above, but to run it very fast, so that CDRs are collected and processed promptly. A number of mobile networks use this approach. However, it has proved difficult to get the billing delay down below about 30 minutes. Therefore when a user has run out of credit, there is a 30-minute window in which calls (as expensive as one likes) can be made before the network bars the phone. Where such networks operate, there is a significant black market in these 30-minute phones.

In market areas where credit management is unattractive either for the telco or for the customers, pre-paid services have become popular.

Another is to use the customer's equipment to store the credit balance. For example, SIM cards have been used in GSM networks in this way. While this offers immediate barring when the credit runs out, it is open to fraud through reprogramming the SIM cards (SIM card programmers are readily available).

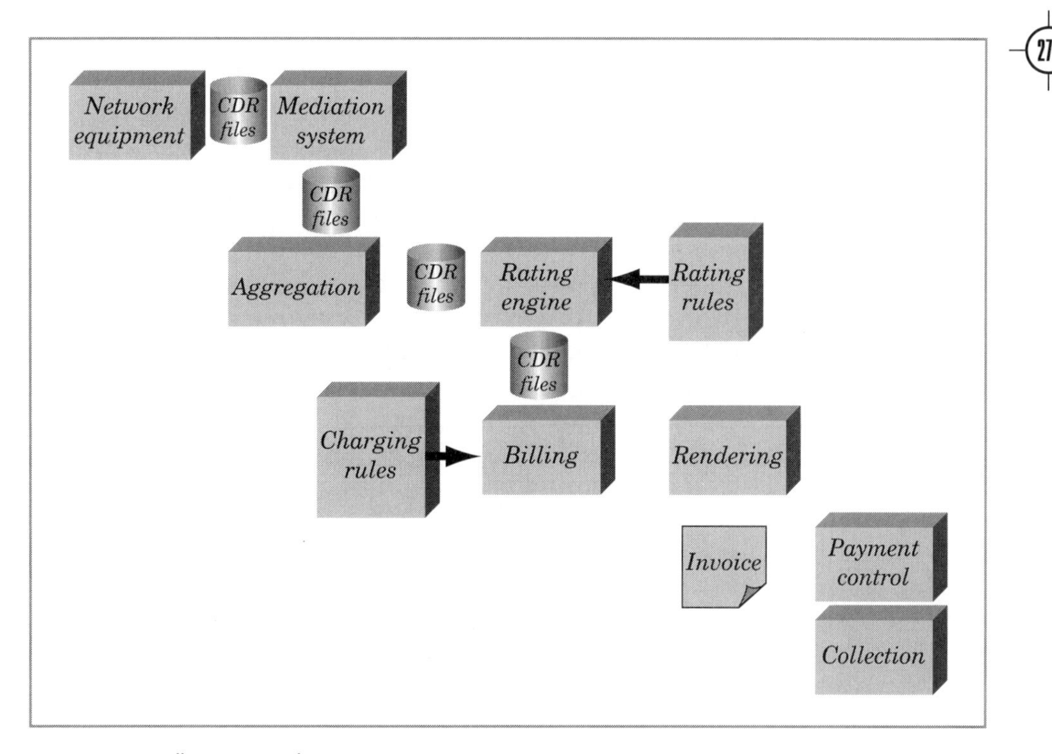

Figure 11.9 Billing system elements

Probably the most common approach is to use IN techniques to intercept the network traffic, refer the call to a credit database and do the barring in real time.

Other BSS

Telco business support systems are in many ways like those in other service industries. However, a few features are peculiar to the telecoms industry. Network and service problems are managed through trouble ticketing systems (also called problem management and fault docket systems). Problems can be reported electronically via the network management infrastructure, or by customers, or by OLOs.[5] Trouble ticketing systems aim to integrate the information from these sources to bring together all the information for each problem, and to track the problem's resolution. The performance of the telco staff in resolving troubles can be another input into service surveillance and SLA automation systems.

Because telecoms networks are highly distributed, geographical information systems are particularly relevant. GIS may be integrated into planning systems, for example for cell site placement. They may be used in sales support systems, for example to link sales campaigns to local network build plans. They may be used to aid diagnosis in trouble management systems, and they may be used in field force management.

The huge volumes of network usage information provided by a network's CDRs can be used as a source for data mining, allowing the customers' behaviour to be analysed in some depth. CDR stores are among the largest data warehouses in the world. Thus when a customer calls their network service provider, the telco operator is presented with a list of new sales propositions, selected for their appeal to that customer's lifestyle.

[5] In some cases, trouble tickets can be exchanged with OLOs electronically. There is an ITU-T recommendation for interoperable trouble tickets, X.790. Also, ad hoc trouble ticket formats can be exchanged called electronic bonding.

Further reading

Adams and Willetts (1996) present a vigorous justification of TMN thinking, as befits their position at the head of the TMF. A starting point for exploring the internet network management standards is RFC 2570 'Introduction to Version 3 of the internet-standard network management framework'. XML is pleasantly introduced in Norris *et al.* (2000).

12 Software and systems issues

The role of software in telecoms systems grew from a negligible part in the 1960s to a position in 2000 where perhaps 85 per cent of network costs were attributable to software.

The software and systems needs of a network operator or telecoms equipment maker are, to an extent, just the same as those of any other business. However, some special characteristics of the telecoms business raise quite unusual systems and software requirements. This chapter introduces those special requirements, then reviews the special software techniques that have been developed to address them.

Telecoms systems issues

Scale and speed

Many telecoms systems have to support large numbers of objects, and process work at a rapid rate. A typical national wireline network might have to support, say, 25 million customers and 26 million local loop connections, and carry 25 million calls per year, across a network of 5000 exchanges. Any system that has to model even a significant fraction of such a network, or participate in even a fraction of those calls, will have to meet challenges of scale and performance.

Availability

In the 1970s, when SPC exchanges first appeared, telecoms network operators became quite unusually dependent on computer systems. At that time, computer systems failures in most businesses could cause the business to be inconvenienced, and even stop some parts of it from operating. But in a telecoms network, a switching systems failure could close down the network, and thus its operator's business. While the rest of the world's dependency on computer systems has grown immensely in the intervening years, telecoms network equipment still requires quite exceptional levels of reliability. For example, service-critical network elements are often required to be continuously available for 20 years, with only a couple of hours' loss of availability in that period (equivalent to 99.999 per cent or 'five-nines' availability). This level of availability has to cover both software and hardware failures, but also the system upgrades that are inevitable in such a period.

In a telecoms network, a switching systems failure could close down the network, and thus its operator's business.

Complexity

Telecoms network devices include some of the most complex programs ever. In particular, SPC circuit switches can be extremely complex, containing anything from 100 000 to 10 million lines of code. The development and support of such immense programs requires large and therefore usually widely distributed teams.

When I worked on one well-known, typical and much-hated SPC switch in the 1980s, there was a team of 800 programmers working on it, split across two continents. The software build was so vast that a rebuild took a week. There were vast swathes of software of unknown function, which nobody dared touch in case they broke something. Things have not got any easier since then.

The scale and complexity of such programs lead to difficulties, in particular with software side effects. These can be at the programming, design or network levels. At the programming level, there are huge numbers of variables and memory blocks to manipulate, and so many lines of code that the risk of some line of code writing to the wrong piece of data becomes high.

At the design level, side effects arise from the difficulty of understanding the whole design well enough to make changes without disrupting the system. It is a typical experience in writing such systems to adjust one aspect of the design and find that another apparently unrelated aspect of the system mysteriously goes wrong.

I once worked on the software team for a PABX, with only about 200 000 lines of code in it. Arthur[1] had designed the 'divert' facility. This had been a real intellectual tour de force because the design had to consider issues such as divert chains, divert loops, divert to hunt group, follow-me and lots more. Arthur's work was coded and tested, and the team moved on to other features. One of those was 'ring back when free'. Bethany, another fine engineer, designed that feature. At a late stage in the work, Bethany realized that ring back would have to take account of the fact that extensions could have divert set. Bethany looked in the system design and found some access routines which would tell her whether this was the case. Bethany's design used them, but failed to take account of the rest of Arthur's work. When ring back was tested on its own, it worked fine. But at system test time, days before the business was expecting to take delivery of the software from the team, someone discovered that divert worked fine, except when in ring back mode. It turned out that Bethany's ideas about divert had been naïve. The way that divert operated in ring back mode was quite different from the way it operated for the rest of the system. Some months were lost in repairing the design.

At least in that case the problem was caught before the system was shipped to hundreds of locations across the globe. With network-level side effects, that is not always possible. Sometimes an individual device can operate perfectly in isolation, or in a small network, but unexpectedly misoperate when put into a large network. For example, in 1999, one IP carrier's network failed for several days. It had installed a new version of its IP router software, which had an undiscovered bug which caused the routers to create routing loops, but only when there were some tens of them connected together. Network-level side effects are often very costly, not only because of their capacity to disrupt live networks but because of the manpower cost of upgrading systems in the field.

Lifetime and change

Many items of telco network equipment are extremely costly, and their capital cost can be covered only if they remain in

[1] These names are altered, to protect the guilty. But was it Bethany's fault, or Arthur's, or the project manager's?

service for 10 or 20 years. This means that the post-acquisition lifetime costs of a system or program frequently exceed the acquisition costs, so software techniques that address the cost of maintenance are in demand. The long software lifespan of these systems also means that programs are often required to span several generations of computing and network technology, which imposes special demands on their adaptability and portability.

As network services change, so do the demands on the software in (and supporting) the network. It is not uncommon for a telco to introduce as many as 30 new service variants per year, and for each service variant to require changes to several network and supporting systems. Competitive pressures mean that often these changes are needed outrageously quickly, considering the complexity of the systems involved.

Rollout

Normally telecoms networks include large numbers of devices. The devices are usually dispersed over a large geographical territory, or even into orbital space. Rolling out a new program version can therefore be a large task. If the devices have to be individually reconfigured to use the new software, or if their upgrades have to be synchronized, the task becomes complex as well.

The risk of a network-level problem appearing after rollout is very serious, because although it is reasonably rare, its consequences are often severe, in terms of lost network service and of network recovery work. Therefore network operators often make the successful network rollout a part of their equipment suppliers' work, which motivates the suppliers to conduct extensive trials before rollout. For example, a new software release for an SPC circuit switch is typically tested in a mockup network of several switches, over a period of many weeks, before rollout into the real network is authorized.

The risk of a network-level problem appearing after rollout is very serious

Connectivity, concurrency and time

Network devices typically have very high peripheral connectivity. A switch may have upwards of 10 000 connections to it. Usually, therefore, the software in the device has to be highly concurrent, often with thousands of threads executing the same program simultaneously. The scheduling problems that this introduces are exacerbated by the close timing tolerances that can be needed. For example, programs in TDM devices often need to be reliably able to service each input stream once every time slot.

Telecoms software methods

The following sections describe some software technologies that have been developed specifically for the telecoms industry. They are:

- a procedural programming language;
- a functional language;
- a finite-state modelling system;
- a data-definition tool;
- a systems toolkit.

This is far from being an exhaustive list of telecoms software technologies, but these items are chosen to illustrate the range available. Notably absent from the list is the Java language which, while it is well adapted for propagation across telecoms networks, is not intended specifically to address telecoms network issues.

CHILL

In the 1970s, when block-structured procedural programming languages were emerging, telecoms manufacturers clubbed together under the auspices of the ITU-T (at that time called the CCITT[2]), to define a block-structured language suitable for programming telecoms network devices. Existing block-structured procedural languages such as Algol, A68, BCPL and Pascal were considered unsuitable because of their lack of explicit support for features such as concurrency. The result was CHILL, the *CCITT high level language*, published in 1980

[2] Commité Consultatif Internationale de Telephonie et Telegraphie.

as CCITT recommendation Z.200 (ISO/IEC 9496).[3] CHILL has been used for a significant fraction of SPC systems worldwide, and is still in widespread use.

As well as the normal features of a block-structured, modular procedural language, CHILL supports:

- explicit control of concurrent execution, including language elements for starting, stopping, delaying threads, for sending and receiving asynchronous messages;
- exception handling, through user definition of error handlers to catch errors;
- time supervision, making information about elapsed and absolute time available as language elements.

Most of these features are also supported by the Ada language, and indeed if Ada had been available earlier, CHILL might never have happened. As CHILL is an unspeakably ugly language, with even fewer enthusiasts than Ada, that might have been a good thing.

Erlang

Erlang is an example of a proprietary telecoms programming language. Erlang was developed by Ericsson, although some use of the language has been made outside that company. A number of other large telecoms equipment manufacturers have developed their own languages; for another example, Nortel has one called Protel. Erlang is particularly interesting as an example because its declarative approach contrasts with most everyday programming languages.

Erlang was of course named after Agner Krarup Erlang, of teletraffic engineering fame. It is a functional or declarative language, meaning that the programs are expressed as mathematical functions, rather than as sequences of steps. So, for example, repetition is achieved in Erlang through recursion, not iteration. Erlang allows no global variables, and does not allow the value of a variable to be changed after its original assignment. Erlang is therefore well suited to formal verification, and also minimizes the possibility of side effects at the programming level.

[3] There have been relatively minor upgrades to the language since then.

Erlang has explicit support for concurrency and timing control, built-in memory management and error handling, facilities for hot swapping of code modules in live systems, and for the transparent distribution of code modules across multiple hardware platforms. Erlang is also capable of producing strikingly compact, terse programs.

SDL

A normal, procedural programming language assumes a model where the computer starts at the beginning of a program, runs through it, collecting inputs when it wants to, and stops at the end. Often in telecoms systems that model is inappropriate. A program in a network node is likely to have to keep running indefinitely, and to react to inputs that appear out of the network suddenly and in no established sequence. This sort of situation is addressed well by the *finite state machine*[4] (FSM) paradigm.

The FSM idea is that each thread of the program alternates between sitting in one of a defined set of *states*, waiting for an input, and making a *transition* between one state and another, in response to some input. Input events are processed one at a time, in the order that they arrive, and they are considered only while the thread is sitting in a state. The simplicity of the FSM model makes it particularly amenable to formal proofs about how a given FSM will behave,[5] as well as suiting the practical requirements of many telecoms systems.

The ITU-T has specified a tool for defining FSMs, the *specification and description language* (SDL), in recommendations Z.100 – Z.104. SDL has two representational forms: the graphical form and the linear form. The two are equivalent, and there are tools on the market which translate between them. The linear form is text-based, and well suited to code development. A number of programs are available to translate between the linear form and various procedural languages. The graphical form is designed for ease of understanding, and can be used to present proposed system behaviour for checking by non-technical people. Figure 12.1 shows a simplified SDL definition for a telephony line card.

This example has four states – idle, dialling, alerting and in

[4] In computer science circles, finite state automata.

[5] For example, Turing's imaginary machine is an FSM.

call. The process defined by this SDL would probably execute in several parallel instances or threads, and each instance could be in a different state at any given time. The process recognizes six input events – offhook, call, digit, remote answer, remote clear and onhook. Each state shows explicitly which events it is

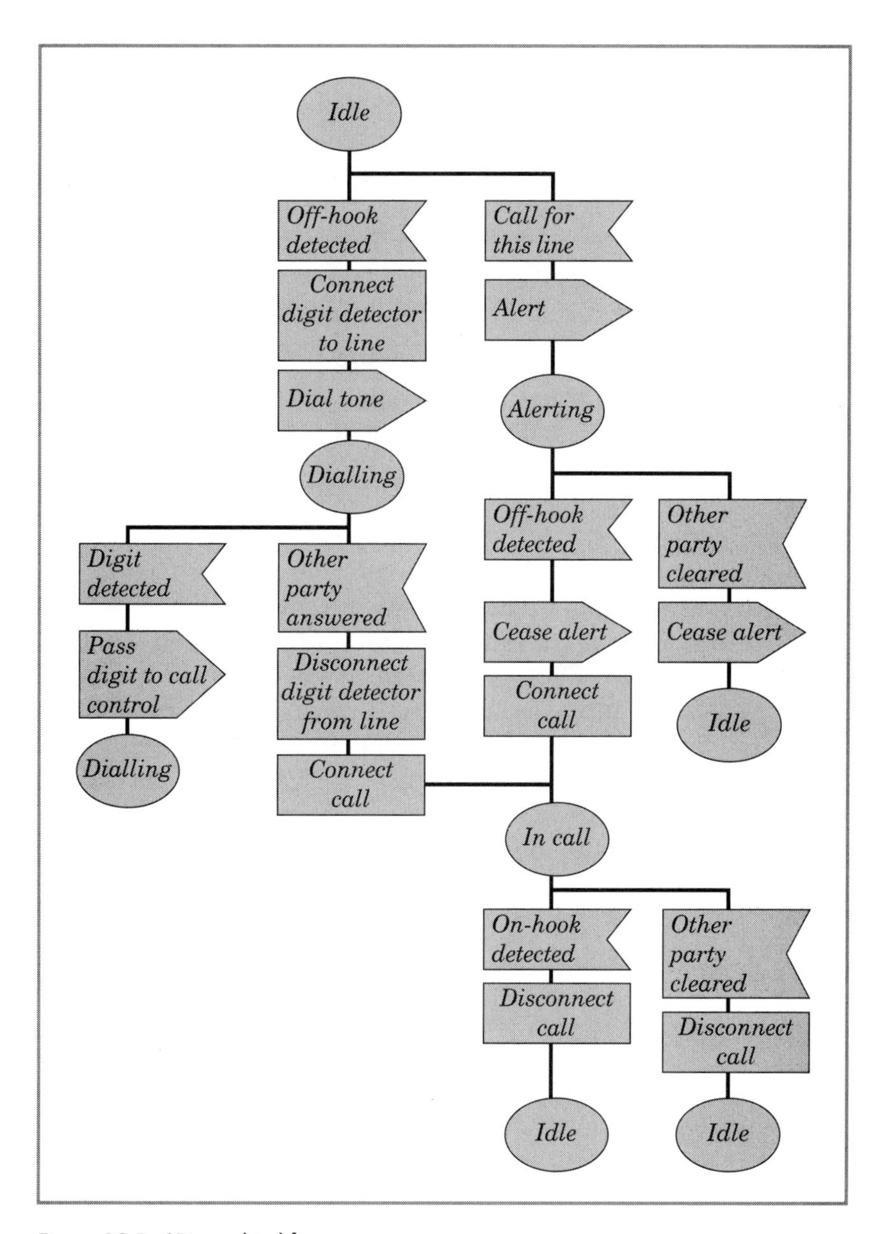

Figure 12.1 SDL graphical form

interested in. Events that are not shown are ignored. So, for example, in state dialling the process is interested in dialled digits. In all the other states, it ignores them. The process can do four tasks – connect digit detector, disconnect digit detector, connect call, disconnect call – and send four output events to other processes – dial tone, alert, digit, cease alert.

Events are assumed to come in down a FIFO queue, so there is no possibility of multiple events arriving simultaneously; that is, the mechanism will make sure that even if they happen simultaneously, the software will see them only one at a time. So in this example, if the system is in the idle state, and the phone user goes offhook just when a call comes in from the network, either:

- the offhook will be processed first, the state will change to dialling, and the incoming call will then be ignored; or
- the incoming call will be processed first, the state will change to alerting, and the offhook will be received and cause immediate transition to 'in call'.

ASN.1

Abstract syntax notation number 1 (ASN.1) is exactly what it calls itself – it is a *notation*, an agreed way of writing things down. It is a notation for *syntax*, that is, for talking about the rules that some set of messages or data structures must obey. It is a notation for *abstract* syntax, because it is concerned with the abstract structure of data. For example, ASN.1 could define a rule saying that IP addresses have to be made of integers separated by full stops; however, ASN.1 could not say anything about how the IP addresses were to be stored or transmitted.[6] Lastly, it is the first (and only) abstract syntax notation that the ITU-T has defined.

Abstract syntax notation number 1 is a notation for abstract syntax, because it is concerned with the abstract structure of data.

ASN.1 is used in many protocol definitions as a way of defining the content and structure of messages without having to worry about the detail of the transfer syntax. It is also used for defining object classes in OMNM, and in various other roles where a widely understood data-definition tool is needed.

[6] Syntaxes which address those sorts of details are called *concrete* syntaxes, and in the case of a transmission format, *transfer* syntaxes.

ITU-T recommendations for ASN.1

X.680 (superseding X.208): Specification of basic notation[7]
X.681: Information object specification
X.682: Constraint specification
X.683: Parameterization of ASN.1 specifications

ASN.1 allows the user to define items, objects, types and values. It provides a palette of built-in data types, but most real-life ASN.1 productions8 include a lot of their own defined types, as the example here illustrates.

ASN.1 example

```
CardStatus ::= CHOICE
{noInformation          [1]      Null,
statusInfo      [2]     ENUMERATED
      {initializing     (1),
      idle              (2),
      busy              (3)}
      }
LineCardConfig ::= SEQUENCE {
      signallingSystem ENUMERATED {loopDis (1), DTMF (2)},
      codec ENUMERATED {muLaw (1), ALaw (2)},
      lineIdentifier PrintableString SIZE (20)
      }
LineCard ::= SET SIZE (4) OF
      {portNumber INTEGER,
      status CardStatus,
      configuration LineCardConfig}
lineCardInformation ::= LineCard
```

[7] There are many tutorial introductions to ASN.1, but X.680 is so lucid that it is easy enough to start from there.
[8] A piece of ASN.1 is called a *production*.

ASN.1 is supported by a number of sets of encoding rules, which get from the abstract syntax of an ASN.1 production to a concrete syntax for use in a real protocol. Each set of rules is designed to address different needs, for example for compression or for simplicity of decoding. ITU-T X.690 (superseding X.209) defines the *basic encoding rules* (BER), *canonical encoding rules* (CER) and *distinguished encoding rules* (DER); X.691 adds the *packed encoding rules* (PER).

ASN.1 covers much the same ground as, say, *Backus Naur form* (BNF) or XML. Indeed, had XML been around when ASN.1 was invented, ASN.1 would have had no justification. However, as it has been around for some time, many standards rely on it, and its use is likely to continue for many years.

Network management frameworks

Software component re-use is nothing new, but one corner of the telecoms industry has found scope for component re-use on a particularly large (and commercially significant) scale. Network management problems are the same in every telco; much the same things need doing, and the equipment to be managed is similar the world over. Yet each telco's business processes are a bit different, and so are their networks, so there is little possibility of putting in a shrink-wrapped network management system. There is therefore a large market for network management frameworks, or network management platforms. There are more than 30 of these on the market, and they sell for between $500 and $5 million. Figure 12.2 illustrates their typical features.

For interfacing with managed devices or lower-order management systems, a range of management protocol stacks is offered. These may be standard (for example, CMIP, SNMP, TL 1 or CORBA), or they may be specific to a device or system (for example, System X configuration management port). As no framework can hope to support all the peculiar management interfaces in the world, most of them include an access toolkit, so that the buyer can build whatever interfaces are needed. At least in principle, all the stacks (and the toolkit) will offer the same API to the rest of the framework, so that the applications need not be aware of what protocol is being used.

Most management frameworks offer distributability across multiple hardware platforms, and most of them also support multiple concurrent users. Distribution is sometimes done by using a standard management protocol such as CMOT, within the framework, but is more likely to be via a generic IT distribution technology such as DCOM, RPC or CORBA.

Above the stacks, there is usually a store for a network model or service model, some standard GUI elements and a number of applications. The framework vendor usually supplies some generic applications (such as geoschematics and an alarm list)[9] and also may offer specific applications for particular problems or network devices (for example, an RMON monitor).

All the framework vendors claim that their offerings are open and extensible. If they really are, they will include an applications

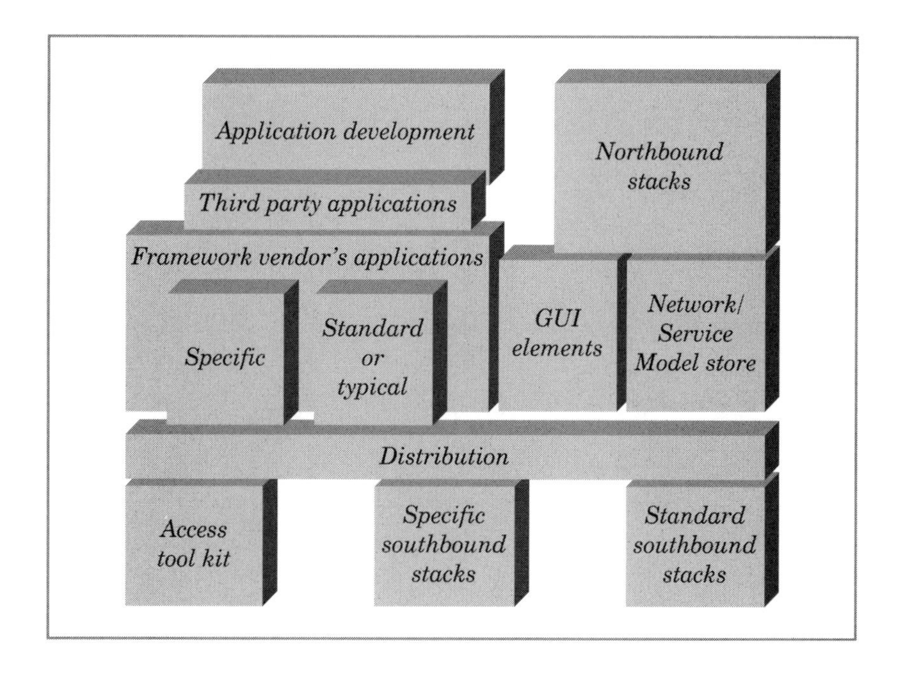

Figure 12.2 Network management framework features

[9] For example, the screens shots shown in Figures 11.6, 11.7 and 11.8 are based on the NetExpert OSS platform offered by Objective Systems Integrators Inc.

development environment, and there will also probably be a range of applications developed by third parties. The absence of such third-party applications is a warning sign that what is being offered as an extensible framework may in fact be an intractable lump of software, which can be modified only by its vendor.

References

1. Adams, E.K. and Willetts, K.J. (1996) *The Lean Communications Provider*, NY: McGraw-Hill.
2. Bateman, A. (1998) *Digital Communications*, Harlow: Addison Wesley Longman.
3. Black, U. (1999) *Advanced* Internet *Technologies*, NJ: Prentice Hall.
4. Brabner, S. and Dellard, C. (1999) 'The Full Monty: PSTN-Equivalent IP Telephony for Spain', *BTEJ*, vol. 18, part 3, Oct, pp. 92–98, London: Institution of BT Engineers.
5. *BT Technology Journal* (1998) 'Local access technologies', vol. 16, no. 4, Dordrecht: Kluwer.
6. Carroll L. (1872) *Through the Looking Glass,* London: Macmillan.
7. Cole M. (1999) *Telecommunications*, NJ: Prentice Hall.
8. Douskalis W. (2000) *IP Telephony*, NJ: Prentice Hall.
9. Fenton, F.M. and Sipes, J.D. (1996) 'Architectural and technological trends in access: an overview', *Bell Labs Technical Journal*, summer, pp. 3–10.
10. Flood, J.E., ed. (1997) *Telecommunication Networks*, Stevenage: Peter Peregrinus.
11. Flood, J.E. and Cochrane, P., eds. (1995) *Transmission Systems*, Stevenage: Peter Peregrinus.
12. Goralski, W.J. (1995) *Introduction to ATM Networking*, NY: McGraw-Hill.

13. Hallows, R. (1995) *Service Management in computing and Telecommunications*, Norwood, MA: Artech House.

14. Halsall, F. (1995) *Data Communications, Computer Networks and Open Systems*, Harlow: Addison Wesley Longman.

15. Hiett, A. (1991) 'Signalling in telecommunications networks', *Structured Information Programme*, 7.1. London: Institution of British Telecommunications Engineers.

16. Isenberg, D. (1997) *The Rise of the Stupid Network*, http://www.hyperorg.com/misc/stupidnet.html (and other sites).

17. Modarressi, A.R. and Skoog, R.A. (1990) 'Signalling system no. 7: a tutorial', *IEEE Communications Magazine*, July, vol. 28, no. 7, pp 19–35.

18. Nilsson, M. (1999) 'Third-generation radio access standards', *Ericsson Review*, no. 3.

19. Norris, M., West, S. and Gaughan, K. (2000) *eBusiness Essentials*, Chichester: John Wiley.

20. Peersman, G. *et al.* (2000) 'A tutorial overview of the short message service within GSM', *IEE Computing & Control Engineering Journal*, April, Stevenage: Peter Peregrinus.

21. Shannon, C.E. (1948) 'A mathematical theory of communication', *Bell System Technical Journal,* vol. 27, no. 3, July.

22. TeleManagement Forum (1999)(a) *Telecom Operations Map*, NJ: TeleManagement Forum.

23. TeleManagement Forum (1999)(b) *Network Management Detailed Operations Map*, NJ: TeleManagement Forum.

24. Wesel, E. (1997) *Wireless Multimedia Communications*, Harlow: Addison Wesley Longman.

Glossary

AAR	automatic alternative routing
ABR	available bit rate
ACD	automatic call distribution
ACE	automatic cross-connect equipment
ACSE	association control service element
ACTA	America's Carriers Telecommunications Association
ADM	add-drop mux
ADSL	asymmetric/asynchronous/advanced digital subscriber loop
Al	aluminium
AM	amplitude modulation
AMA	automatic management accounting
AMPS	advanced mobile phone service
AMR	adaptive multirate
AN	access network
ANI	access network interface
ANPS	advanced network protection system
ANSI	American National Standards Institute
API	application programming interface
ARIB	Association of Radio Industries and Businesses
ARP	address resolution protocol
ASCCH	assigned secondary control channel
ASCII	American standard code for information infiltration
ASE	application service element
ASK	amplitude shift keying
ASN	abstract syntax notation
ATM	asynchronous transfer mode
ATMF	ATM forum
ATT	automatic toll ticketing

AuC	authentication centre
BABT	British Board of Approvals for Telecommunications
BER	basic encoding rules
BGP	border gateway protocol
B-ISDN	broadband ISDN
BOOTP	bootstrap protocol
bps	bits per second
BRI	basic rate ISDN
BSC	base station controller
BSS	business support system
BTS	base transceiver station
C6	(common channel) signalling system number 6
C7	(common channel) signalling system number 7 (= SS7)
CA	conditional access
Camel	customized applications for mobile enhanced logic
CAS	channel associated signalling
CATV	community antenna TV (etc.)
CB	central battery (abbreviation used in CAS)
CB	citizens' band (abbreviation used in radio spectrum)
CBDS	connectionless broadband data service
CBR	constant (or continuous) bit rate
CBS	central battery signalling (abbreviation used in CAS)
CBS	common base station (abbreviation used in PAMR)
CBX	computerized branch exchange
CCS	common channel signalling
CDMA	code division multiple access
CDPD	cellular digital packet data
CDR	call detail record
CDSL	consumer DSL
CEM	computing equipment vendor
CEPT	Conference Européen Des Postes et Telecommunication
CGI	common gateway interface
CHILL	CCITT high level language
CIDR	classless inter-domain routing
CIR	committed information rate
CLASS	custom local area signalling services

CLEC competitive local exchange carrier
CLI calling line identification (or identity)
CLIP calling line identity presentation
CLIR calling line identity restriction
CMIP common management information protocol
CMIS common management information service
CMISE common management information service element
CMM capability maturity model
CMOT CMIP over TCP/IP
CMTS cable modem termination system
CNM customer network management
COPS common open policy service
CP concentration point
CPE customer premises equipment
cps chips per second (abbreviation used in CDMA)
CRC cyclic redundant checksum
CS convergence sublayer (abbreviation used in ATM)
CS-1 capability set 1 (abbreviation used in IN)
CSD circuit switched data (abbreviation used in GSM)
CTI computer telephony integration
Cu copper
CUG closed user group

DACCS digital access cross-connect system
DACS digital access carrier system
DASS digital automatic (or access) signalling system
DAWS digital advanced wireless service
dB decibels
dBm decibels relative to one milliwatt
DC direct current
DCE data communications equipment
DCS digital communication system
DDD direct distance dialling
DDI direct dial in
DECT digital enhanced cordless telephony
DF distribution frame
DHCP dynamic host configuration protocol
DID direct inbound dialling
DISA direct inward system access
DMO direct mode operation
DN distinguished name
DNIC data network identification code

DNS	domain name system
DOCSIS	Data Over Cable Service Interface Specification
DoD	Department of Defense
DOD	direct outward dial
DP	distribution point
DPC	destination point code
DPNSS	digital private network signalling system
DQ	directory enquiries
DS	derived services
DSL	digital subscriber line
DSLAM	digital subscriber line access multiplexer
DSS1	digital subscriber signalling 1
DTE	data terminal equipment
DTMF	dual tone multi-frequency
DTX	discontinuous transmission
DUP	data user part
DWDM	dense wavelength division multiplexing
E	Erlang
ECMA	European Computer Manufacturers' Association
EDFA	erbium doped fibre amplifier
EDGE	enhanced data rate for GSM (or global) evolution
EEPROM	electrically erasable programmable read-only memory
EGP	exterior gateway protocol
EMS	element management system
ERAN	EDGE radio access network
ERMES	European radio messaging system
ESN	electronic serial number (TACS)
ET	exchange termination
ETS	European telecommunications standard
ETSI	European Telecommunications Standards Institute
EU	European Union
FCC	Federal Communications Commission
FDD	frequency division duplexing
FDM	frequency division multiplexing
FDMA	frequency division multiple access
FIPS	Federal Information Processing Standards
FISU	fill-in signal unit
FM	frequency modulation
FoIP	fax over IP

FPLMTS	Future Public Land Mobile Telephony System
FRA	fixed radio access
FSAN	full service access network
FSK	frequency shift keying
FSM	finite state machine
FT	fault tolerant
FTAM	file transfer and manipulation protocol
FTP	file transfer protocol
FTTB	fibre to the building
FTTC	fibre to the curb
FTTCab	fibre to the cabinet
FTTH	fibre to the home
FTTK	fibre to the kerb
GCRA	generic cell rate algorithm
GDMO	Guidelines for the definition of managed objects
GEO	geostationary earth orbit
GGSN	gateway gprs support node
GIS	geographical information system
GMPCS	global mobile personal communications by satellite
GPO	General Post Office
GPRS	general packet radio service
GPS	global positioning system
GSM	global system for mobile
GSN	GPRS support node
GUI	graphical user interface
HAP	high-altitude platform
HDLC	high-level data link control
HDML	handheld device markup language
HDSL	high-speed digital subscriber loop
HDTV	high definition TV
HFC	hybrid fibre co-ax
HLR	home location register
HMMP	HyperMedia Management Protocol
HMMS	HyperMedia Management Schema
HMOM	HyperMedia Object Manager
HSCSD	high-speed circuit switched data
HTML	hypertext markup language
Hz	hertz
IAB	Internet Architecture Board

IANA	Internet Assigned Numbers Authority
ICANN	Internet Corporation for Assigned Names and Numbers
ICLID	individual calling line ID
ICMP	Internet control message protocol
IDSL	ISDN DSL
IEC	inter-exchange carrier
IESG	Internet Engineering Steering Group
IETF	Internet Engineering Task Force
IGMP	Internet group management protocol
IGP	interior gateway protocol
ILEC	independent local exchange carrier
IMAP	Internet message access protocol
IMT2000	International Mobile Telecommunications 2000
IMTC	Internet Multimedia Technical Committee
IN	intelligent networks
INF	IN forum
Internic	Internet network information centre
IP	intelligent peripheral (abbreviation used in IN)
IP	Internet (or Internetwork) protocol
IPDC	Internet protocol device control
IPng	IP next generation (= IPv6)
IPPV	impulse pay per view
IPR	intellectual property rights
IRD	integrated receiver/decoder
IrDA	Infrared Data Association
ISDN	integrated services digital network
ISO	International Organization for Standardization
ISOC	Internet Society
ISP	Internet service provider
ISR	international simple resale
ISV	independent software vendor
ITA	international telegraphy alphabet
ITU	International Telecommunications Union
IVR	interactive voice response
JAIN	Java advanced IN
Janet	Joint Academic Network
JCB	(initials of a well-known manufacturer) excavator
JDMK	Java dynamic management kit
JTAPI	Java telephony API

L2F	Layer 2 forwarding
L2TP	Layer 2 tunnelling protocol
LAC	L2TP access concentrator
LAN	local area network
LANE	LAN emulation
LAP	link access protocol
LAP-B	link access protocol, balanced
LAPD	link access protocol, D-channel
LASS	Local Area Signalling Services
LCP	link control protocol
LDAP	lightweight directory access protocol
LE	local exchange
LEC	local exchange carrier
LED	light emitting diode
LEO	low earth orbit
LMDS	local multipoint distribution service
LNS	L2TP network server
LPC	linear predictive coding
LQR	link quality report
LSSU	link status signal unit
LT	line termination

MAP	mobile application part
Mbone	multicast backbone
MBS	maximum burst size (abbreviation used in ATM)
MBS	mobile broadband system (abbreviation used in fourth generation mobile systems)
MCU	multipoint control unit
MDF	main distribution frame
MEO	medium earth orbit
MexE	mobile station application execution environment
MF	mediation function (abbreviation used in TMN)
MF	multi-frequency (abbreviation used in circuit mode signalling)
MGCP	media gateway control protocol
MIB	management information base
MIME	multipurpose Internet mail extensions
MIN	mobile identification number (TACS)
MoU	Memorandum of Understanding
MPEG	Motion Picture Experts Group
MPLS	multiprotocol label switching
MRVT	MTP routing verification test

MS	mobile station
MSC	mobile switching centre
MSISDN	mobile subscriber ISDN number
MSO	multiple service operator
MSU	message signal unit
MTP	message transfer part
NAMPS	narrowband AMPS
NBS	National Bureau of Standards
NCS	network-based call signalling
NE	network element
NEF	network element function
N-ISDN	narrowband ISDN
NLMS	network level management system
NMS	network management system
NMT	Nordic mobile telephony
NNI	network/network interface
NNTP	Network News Transfer Protocol
NOC	network operations centre
NPA	network point of attachment
NSAP	network service access point
NT	network termination
NTE	network termination equipment
NTP	network time protocol
NTT	Nippon Telegraph & Telephone Corp.
OAN	optical access networking
OLO	other licensed operator
OLT	optical line terminator (e.g. for PON)
OLTE	optical line termination equipment
OMAP	operations maintenance and administration part
OMNM	object modelled network management
ONT	optical network termination
ONU	optical network unit (e.g. for PON)
OOS	out of service
OPC	originating point code
OPSF	open path shortest first
OSF	operations system function
OSI	open systems interconnection
OSS	operations support systems
PABX	private automatic branch exchange

PACS	personal access communications system
PAD	packet assembler and disassembler
PAMR	public access mobile radio
PANS	pretty amazing (or awesome) new stuff
PBX	private branch exchange
PCM	pulse code modulation
PCN	personal communications network
PCR	peak cell rate (abbreviation used in ATM)
PCR	preventative cycle retransmission (abbreviation used in SS7)
PCS	personal communications system
PCU	packet control unit
PDC	personal digital cellular
PDH	plesiochronous digital hierarchy
PDP	policy decision point
PDU	protocol data unit
PEP	policy enforcement point
PHS	personal handyphone system
PLP	packet level protocol
PMR	private (or professional) mobile radio
POCSAG	Post Office Code Standardization Advisory Group
PON	passive optical network
POP	point of presence (abbreviation used in network architectures)
POP	post office protocol (abbreviation used in email handling)
POTS	plain old telephony service
PPM	periodic pulse metering
PPM	pulse position modulation
PPP	point-to-point protocol
PPS	pulses per second
PPTP	point-to-point tunnelling protocol
PPV	pay per view
PRI	primary rate ISDN
PSK	phase shift keying
PSS	packet switched service
PSTN	public switched telephony network
PTO	posts and telecoms operator
PTT	posts, telephony and telegraphy operator
PVC	permanent virtual circuit

Q.Sig (ETSI)	Unified international digital corporate network signalling standard
QAF	Q-adaption function
QAM	quadrature amplitude modulation
QoS	quality of service
QPSK	quadrature phase shift keying
R1	regional signalling system number 1
RADSL	rate-adaptive DSL
RAM	random access memory
RARP	reverse address resolution protocol
RBOC	regional Bell operating company
RCU	remote concentrator unit
RDN	relative distinguished name
RF	radio frequency
RFC	request for comments
RFI	radio frequency interface
RIP	routing information protocol
RITL	radio in the loop
RLL	radio local loop
RMON	remote network monitoring
RNC	radio network controller
ROSE	remote operations service element
RPE-LPC	regular pulse excited linear prediction coding
RSVP	resource reservation protocol
RTCP	real-time control protocol
RTNR	real-time network routing
RTP	real-time protocol
RTT	radio transmission technology
SAAL	signalling ATM adaption layer
SAR	segmentation and reassembly
SCCP	signalling connection control part
SCE	service creation environment
SCF	selective call forwarding (Class feature)
SCF	service control function (abbreviation used in IN architectures)
SCP	service control point
SCR	selective call rejection (abbreviation used in IN)
SCR	sustainable bit rate (abbreviation used in ATM)
SDH	synchronous digital hierarchy
SDL	specification and description language

SDS	short data service
SDSL	single line DSL
SEAL	simple and efficient adaptation layer
SEI	Software Engineering Institute
SGCP	simple gateway control protocol
SGML	standard generalized markup language
SGSN	serving GPRS support node
SI	systems integrator
SIB	service independent building block
SIM	subscriber identity module
SIP	session initiation protocol
SLA	service level agreement
SLEE	service logic execution environment
SLG	service level guarantee
SLP	service logic program
SLR	service location register
SLS	signalling link selection
SMASE	system management application service element
SMDS	switched multimegabit data service
SMG	Special Mobile Group
SMI	structure of management information
SMON	switch monitoring
SMS	service management system (abbreviation used in IN)
SMS	short message service (abbreviation used in GSM)
SMSC	short message service centre
SMTP	simple mail transfer protocol
SNI	service node interface
SNMP	simple network management protocol
SPC	source point code (abbreviation used in SS7)
SPC	Stored Program Control (abbreviation used for circuit mode switches)
Spring	shared protection ring
SS6	signalling system number 6
SS7	signalling system number 7
SSCOP	service-specific connection-oriented protocol
SSP	service switching point
STD	subscriber trunk dialling
STM	synchronous transport module
STP	signal transfer point
STS	space time space
STS	synchronous transport signal

SVC	switched virtual circuit
TA	terminal adapter
TACS	total access communications system
TAPI	telephony API
TCAP	transaction capabilities application part
TCP	transmission control protocol
TDD	time division duplexing
TDM	time division multiplexing
TDMA	time division multiple access
TE	terminal equipment
TEM	telecoms equipment vendor
TETRA	trans-European trunked radio access or terrestrial trunked radio
TFTP	trivial file transfer protocol
TINA	telecommunications information networking architecture
TL/1	transaction language number 1
TMF	TeleManagement Forum
TMN	telecommunications management network
TNPP	telelocator network paging protocol
TOM	telecom operations map
TPON	telecommunications over passive optical networks (aka PON)
TRAU	transcoder rate adaption unit
TSAPI	telephony services API
TST	time space time
TUP	telephony user part
UBR	unspecified bit rate
UCD	uniform call distribution
UDP	user datagram protocol
UDSL	unidirectional DSL
UML	universal markup language
UMTS	universal mobile telecommunications system
UN	United Nations
UNI	user/network interface
URL	uniform resource locator
USSD	unstructured supplementary services data
UTRA	UMTS terrestrial radio access
UTRAN	UMTS terrestrial radio access network
UWCC	Universal Wireless Communications Consortium

VBR	variable bit rate
VC	virtual circuit (or virtual channel in ATM)
VC	virtual container (abbreviation used in SDH)
VDSL	very high-rate ADSL
VLR	visitor location register
VoIP	voice over IP
VON	voice over the net
VP	virtual path
VPN	virtual private network
VRML	virtual reality markup language
VSAT	very small aperture
W3C	WWW consortium
WAE	wireless application environment
WAP	wireless application protocol
WARC	World Administrative Radio Conference
WBEM	Web-Based Enterprise Management Forum
WCDMA	wideband CDMA
WDM	wavelength division multiplexing
WDP	wireless datagram protocol
WF	workstation function
WLANA	Wireless LAN Alliance
WLL	wireless local loop
WML	wireless markup language
WSP	wireless session protocol
WTA	wireless telephony application
WTLS	wireless transport layer security
WTP	wireless transaction protocol
WWW	world wide web
XML	extensible markup language

Index

309

312

313

314

317